PREFACE

The present Lecture Notes evolved from a course given at the Technische Hogeschool Eindhoven and later at the Technische Hogeschool Twente. They are intended for computer science students; more specifically, their goal is to introduce the notions of computability and decidability, and to prepare for the study of automata theory, formal language theory and the theory of computing. Except for a general mathematical background no preliminary knowledge is presupposed, but some experience in programming may be helpful.

While classical treatises on computability and decidability are oriented towards the foundation of mathematics or mathematical logic, the present notes try to relate the subject to computer science. Therefore, the exposé is based on the use of strings rather than on that of natural numbers; the notations are similar to those in use in automata theory; in addition, according to a common usage in formal language theory, most of the proofs of computability are reduced to the semi-formal description of a procedure the constructivity of which is apparent to anybody having some programming experience.

Notwithstanding these facts the subject is treated with mathematical rigor; a great number of informal comments are inserted in order to allow a good intuitive understanding.

I am indebted to all those who drew my attention to some errors and ambiguities in a preliminary version of these Notes.

I want also to thank Miss L.A. Krukerink for her diligence in typing the manuscript.

Saarbrücken, March 1972 Jacques Loeckx

TO THE READER

A list of the different notations used throughout the text may be found in Appendix II.

An alphabetic list of the mathematical concepts introduced is in Appendix III.

The end of each example is marked by the sign ** in order to clearly separate it from the running text.

Lecture Notes in Economics and Mathematical Systems

Operations Research, Computer Science, Social Science

Edited by M. Beckmann, Providence, G. Goos, Karlsruhe, and H. P. Künzi, Zürich

68

J. Loeckx

Computability and Decidability

An Introduction for Students of Computer Science

Springer-Verlag
Berlin · Heidelberg · New York 1972

Prof. Dr. Jacques Loeckx
Institut für Informatik
Universität des Saarlandes
66 Saarbrücken

AMS Subject Classifications (1970): Primary 68-01
Secondary 02 E 10, 02 E 15, 02 F 15, 02 G 05, 68 A 05, 68 A 10, 68 A 25, 68 A 30

ISBN 3-540-05869-9 Springer-Verlag Berlin Heidelberg New York
ISBN 0-387-05869-9 Springer-Verlag New York Heidelberg Berlin.

Offsetdruck: Julius Beltz, Hemsbach/Bergstr.

CONTENTS

CHAPTER 1 : SETS AND FUNCTIONS

Abstract: Some notions and definitions related to sets and functions are briefly recalled.

1.1. The objects

This course deals with sets and functions of objects. While in Sections 1.2 to 1.4 the exact nature of the objects is not yet specified, only three different types of objects occur in these Notes:

(i) strings of symbols, such as:

abaa

a

a+b*c **

(ii) non-negative integers or, abbreviated, nni's such as:

7

4013

0 **

(iii) the truth values, viz. true and false.

Actually, the notions of computability and decidability will be defined for functions and sets of symbol strings; functions and sets of nni's or truth values will therefore occur only incidentally.

1.2. Ordered sequences and sets

1.2.1. Definitions

An *ordered sequence (of n objects)* (or: *n-tuple*) consists of a number (viz. n) of objects specified in a certain order; this number may not be zero and must be finite. A 2-tuple and 3-tuple are also called *pair* and *triple* respectively.

A *set* consists of a collection of *elements*, each element being

- an object
- or: an ordered sequence
- or: (another) set (*);

the number of these elements may be zero and must not be finite.

It should be noted that these "definitions" are intuitive descriptions

---

(*) This definition is recursive: the notion to be defined (viz. "set") appears in its own definition.

rather than actual definitions because notions such as "finite" or "collection" have not been defined. Giving precise definitions for these notions is beyond the scope of these Notes (*).

1.2.2. Notation

1.2.2.1. Ordered sequence

An ordered sequence is noted:

(objects separated by comma's)

Examples of ordered sequences are

(3, 9)

(b, a, b)

(a, ab, ba, true) **

Counterexamples of ordered sequences are

( )

(0, 1, 4, 9, ...) where the dots represent an infinite sequence **

Note that an object may occur more than once in an ordered sequence, for instance b in (b, a, b).

Note also that by its definition an ordered sequence of one object is the object itself; in that case the embracing parentheses may, of course, be omitted in the notation.

1.2.2.2. Sets

There exist two different notations.

The first notation is:

{ the elements separated by comma's }

Examples are

{ 3, 9 }

{ a, b, c }

{ } (**)

{ (a, b) , (b, c) , (c, a) }

{ {} , { a } , { b } , { a, b } } **

Note that an element may be mentioned only once; hence

{ a, b, a }

is an incorrect notation for { a, b } (or, equivalently, for { b, a }).

(*) Precise definitions should, among others, avoid the Russell paradox for sets.

(**) This set is called the *empty set*.

The second notation is:

{ variable | property of that variable }

The elements of the set are those satisfying the property (*). Examples are:

{ (x, y) | x and y are nni's for which $x \geq y$ }

{ x | there exists an nni y such that $x = y^2$ } **

An important difference between these two notations is that the first one can only be used for sets with a finite number of elements whereas the second can be used for arbitrary sets.

It should be noted that by its very definition a set of one element is different from the element itself. For instance, {b} denotes a set of one element whereas b denotes an object (or, equivalently, an ordered sequence of one object).

1.3. Further notations and definitions concerning sets

1.3.1. General

$a \in S$ means that a is an element of the set S.

$a \notin S$ means that a is not an element of the set S.

$S = T$ means that $a \in S$ implies $a \in T$ and vice versa (**).

$S \neq T$ means that not $S = T$ (**).

$S \subseteq T$ means that $a \in S$ implies $a \in T$.

$S \subset T$ means that $S \subseteq T$ and $S \neq T$.

A set S is a *subset* of a set T if $S \subseteq T$.

A set S is a *proper subset* of a set T if $S \subset T$.

Two sets S and T are *disjoint* if $a \in S$ implies $a \notin T$.

A set is *finite* if it consists of a finite number of elements.

In the sequel card (S) denotes the number of elements of the set S provided S is a finite set (***); note that card (S) is an nni.

(*) It is understood that the variable is ranging over the elements of the set to be defined.

(**) and, or and not are to be understood as the logical and, or and not (sometimes denoted as $\wedge$, $\vee$, $\neg$ respectively).

(***) The mnemonic origin of "card" is cardinal number.

1.3.2. Constructions on sets

Per definition:

(i) $\underline{S} \cup \underline{T} = \{ a \mid a \in \underline{S}$ or $a \in \underline{T} \}$ (*); $\underline{S} \cup \underline{T}$ is called the *union* of the sets $\underline{S}$ and $\underline{T}$;

(ii) $\underline{S} \cap \underline{T} = \{ a \mid a \in \underline{S}$ and $a \in \underline{T} \}$; $\underline{S} \cap \underline{T}$ is called the *intersection* of the sets $\underline{S}$ and $\underline{T}$;

(iii) $\underline{S} \setminus \underline{T} = \{ a \mid a \in \underline{S}$ and $a \notin \underline{T} \}$; $\underline{S} \setminus \underline{T}$ is called the *difference* of the sets $\underline{S}$ and $\underline{T}$;

(iv) if $\underline{T}$ is a subset of $\underline{S}$, $\underline{S} - \underline{T} = \underline{S} \setminus \underline{T}$ (**); $\underline{S} - \underline{T}$ is called the *complement* of $\underline{T}$ with respect to $\underline{S}$; if $\underline{S}$ is understood, $\underline{S} - \underline{T}$ may be written $\overline{\underline{T}}$;

(v) if $\underline{S}_1, \underline{S}_2, \ldots, \underline{S}_n$ $(n > 1)$ are sets the elements of which are objects, then

$$\underline{S}_1 \times \underline{S}_2 \times \ldots \times \underline{S}_n = \{(a_1, a_2, \ldots, a_n) \mid a_i \in \underline{S}_i \text{ for each } i, 1 \leq i \leq n\} ;$$

this set is called the *cartesian product* of the sets $\underline{S}_1, \underline{S}_2, \ldots, \underline{S}_n$; when $\underline{S}_1 = \underline{S}_2 = \ldots = \underline{S}_n = \underline{S}$, the cartesian product may be written $\underline{S}^n$.

These definitions may be illustrated by the following examples in which $\underline{A} = \{ a, b \}$ and $\underline{B} = \{ b, c \}$:

$\underline{A} \cup \underline{B} = \{ a, b, c \}$

$\underline{A} \cap \underline{B} = \{ b \}$

$\underline{A} \setminus \underline{B} = \{ a \}$

$\underline{A} - \underline{B}$ makes no sense

$\underline{A} \times \underline{B} = \{(a, b), (a, c), (b, b), (b, c)\}$

**

1.4. Functions

1.4.1. Definition

Let $\underline{S}$ be the set of all objects considered (e.g. the set of all nni's).

A *function with n arguments* (or: *n-ary function*, or: *n-ary mapping*, or *n-ary operation*) is a subset of $\underline{S}^{n+1}$ such that for any two different elements

---

(*) and, or and not are to be understood as the logical and, or and not (sometimes denoted as $\wedge$, $\vee$, $\neg$ respectively)

(**) If $\underline{T}$ is not a subset of $\underline{S}$, $\underline{S} - \underline{T}$ makes no sense.

$$(a_1, a_2, \ldots, a_{n+1})$$

and

$$(b_1, b_2, \ldots, b_{n+1})$$

we have not simultaneously $a_1=b_1$, $a_2=b_2, \ldots$, $a_n=b_n$ and $a_{n+1} \neq b_{n+1}$.

If $(a_1, a_2, \ldots, a_{n+1})$ is an element of an n-ary function then a_{n+1} is called *the value (of the function for the n-tuple* $(a_1, a_2, \ldots, a_n)$ *of arguments).*

If there exists no a_{n+1} such that $(a_1, a_2, \ldots, a_n, a_{n+1})$ is an element of the function, then *the value (of the function for the n-tuple* $(a_1, a_2, \ldots, a_n)$ *of arguments) is undefined.*

The following three examples illustrate these definitions:

(i) the function sum is a set containing the elements:

(0, 0, 0), (1, 0, 1), (2, 0, 2), ...
(0, 1, 1), (1, 1, 2), (2, 1, 3), ...
(0, 2, 2), (1, 2, 3), (2, 2, 4), ...
.
.
.

(ii) the function sqrt is a set containing the elements:

(0,0), (1,1), (4,2), (9,3), ... ;

the value of this function is, for instance, 3 for the argument 9 and undefined for the argument 7.

(iii) the 2-ary function dec defined by $10x + y$ for $0 \leq x \leq 9$, $0 \leq y \leq 9$ is the finite set:

dec = { (0, 0, 0), (0, 1, 1), ..., (9, 9, 99) }. **

1.4.2. Domain, range

The *domain* of an n-ary function F is the set

$$\underline{D}_F = \{(a_1, a_2, \ldots, a_n) \mid \text{there exists } a_{n+1} \text{ such that } (a_1, a_2, \ldots, a_n, a_{n+1}) \in F\} \quad ;$$

the *range* of this function is

$$\underline{R}_F = \{a_{n+1} \mid \text{there exist } a_1, a_2, \ldots, a_n \text{ such that } (a_1, a_2, \ldots, a_n, a_{n+1}) \in F\}.$$

One says that F maps $\underline{D}_F$ *onto* $\underline{R}_F$. One also says that F maps $\underline{D}_F$ *into* $\underline{R}$ where $\underline{R}$ is any set such that $\underline{R}_F \subseteq \underline{R}$.

If we consider the functions sum and sqrt of Section 1.4.1 and if $\underline{N}$ denotes the set of all nni's:

$$\underline{D}_{sum} = \underline{N}^2$$

$$\underline{R}_{sum} = \underline{R}_{sqrt} = \underline{N}$$

$$\underline{D}_{sqrt} = \{ x \mid \text{there exists } y \in \underline{N} \text{ such that } x = y^2 \}$$ **

Note that the value of a function for a given n-tuple of arguments is defined if and only if this n-tuple is an element of the domain.

1.4.3. Notation

A function is denoted, for instance, by F or sqrt or sin. The value of the function F for the n-tuple $(a_1, a_2, \ldots, a_n)$ of arguments is denoted by $F(a_1, a_2, \ldots, a_n)$. Note that this is different from the common mathematical practice where sin x denotes the function sine as well as the value of this function for the argument x, the ambiguity in the notation being removed by the context.

The fact that F maps $\underline{S}$ into $\underline{T}$ is noted

$$F : \underline{S} \rightarrow \underline{T}$$;

the fact that F maps $\underline{S}$ onto $\underline{T}$ is noted

$$F : \underline{S} \twoheadrightarrow \underline{T}$$

The definition of a particular function may be given in two ways:

(i) the function may be defined as a set;
examples:

$$F_1 = \{ (x, y, z) \mid x, y, z \text{ are nni's such that } z = x + y \}$$

$$F_2 = \{ (x, 1) \mid \text{there exists an nni } z \text{ such that } x = z^2 \}$$
$$\cup \{ (x, 0) \mid \text{there exists no nni } z \text{ such that } x = z^2 \}$$ **

(ii) the function may be defined with the help of its value according to the scheme

$$F : \underline{D}_F \twoheadrightarrow \underline{R}_F : F(x_1, \ldots, x_n) = \ldots$$

or

$$F : \underline{D}_F \rightarrow \underline{R} : F(x_1, \ldots, x_n) = \ldots$$

where $\underline{R}$ is any set such that $\underline{R}_F \subset \underline{R}$;

examples (*):

$F_1 : \underline{N}^2 \rightarrow \underline{N} : F_1(x, y) = x + y$

$F_2 : \underline{N} \rightarrow \{ 0, 1 \} : F_2(x) = \begin{cases} 1 \text{ if there exists } z \in \underline{N} \text{ such that } x = z^2 \\ 0 \text{ if there exists no } z \in \underline{N} \text{ such that } x = z^2 \end{cases}$

$F_3 : \underline{N} - \{ 2 \} \rightarrow \{ 0 \} : F_3(x) = 0$ (the value of this function is 0 for any argument except for 2; in that case the value of the function is undefined). **

1.4.4. A possible alternative definition for function

A drawback of the definition of a function as given above is that it does in general not tell how to obtain the value for a given n-tuple of arguments. For example, the definition of F_2 in Section 1.4.3 does not tell how to obtain its value for a given x.

Hence a "better" definition would be the following: an n-ary function is a rule which makes correspond with each element of a set of n-tuples (the domain) a unique element of another set (the range). Such a definition nevertheless requires a precise definition of the notion "rule".

It is precisely one of the aims of these Notes to establish such a definition of "rule". In Chapter 5 it will then appear that the two definitions of function are not equivalent: the present definition (i.e. the one of Section 1.4.1) is broader (i.e. defines more functions) than the definition given with the help of the rule.

1.4.5. Injective function and inverse function

An n-ary function is *injective* if for any two different elements

$$(a_1, a_2, \ldots, a_n, a_{n+1})$$

and

$$(b_1, b_2, \ldots, b_n, b_{n+1})$$

it is not the case that $a_{n+1} = b_{n+1}$. Hence for an injective function there corresponds a unique element of the domain with each element of the range.

Examples of injective functions are the functions sqrt and dec of Section 1.4.1; a counterexample is the function sum of the same section. **

(*) $\underline{N}$ denotes the set of all nni's.

The *inverse function of an injective 1-ary function* F is the function

$$\overline{F} = \{ (y, x) \mid (x, y) \in F \}$$

Note that the domain and range of $\overline{F}$ are respectively the range and domain of F. The inverse function is also a 1-ary injective function; in fact, $\overline{\overline{F}} = F$.

As an example, the inverse function of the function "square root of an nni" is the function "square of an nni". **

The *inverse functions of an injective n-ary function* F are the n functions

$$\overline{F}^{(i)} = \{ (y, x_i) \mid \text{there exist } x_1, x_2, \ldots, x_{i-1}, x_{i+1}, \ldots, x_n \text{ such that } (x_1, x_2, \ldots, x_n, y) \in F \} (1 \leq i \leq n, n \geq 2)$$

Less formally, the inverse function $\overline{F}^{(i)}$ restitutes the $i^{\underline{th}}$ argument of the function F for a given value of this function. Hence, for any element y of the range of an injective n-ary function F it is the case that

$$F(\overline{F}^{(1)}(y), \overline{F}^{(2)}(y), \ldots, \overline{F}^{(n)}(y)) = y$$

As an example, the value of the inverse functions $\overline{dec}^{(1)}$ and $\overline{dec}^{(2)}$ of the function dec of Section 1.4.1 is illustrated by

$$\overline{dec}^{(1)}(49) = 4, \overline{dec}^{(2)}(49) = 9$$ **

1.4.6. Note

It is possible to generalize the notion of function by allowing the arguments and the value to be ordered sequences (of objects) and/or sets (of objects) instead of merely objects. In these Notes this generalization will not be made, except for a very particular case mentioned in Section 1.5.1.

1.5. Particular objects

The preceding definitions have been given for arbitrary objects. Particular types of objects or, more precisely, sets the elements of which are constituted by objects of a particular nature, viz. truth values, nni's and strings, are now going to be considered.

1.5.1. Truth values

The truth values are true and false.

A *predicate* is a function the range of which is (a subset of) the set { true, false }.

A *relation* is a 2-ary predicate; example (*):

$$\text{greater}: \underline{N}^2 \rightarrow \{\underline{\text{true}}, \underline{\text{false}}\}:$$

$$\text{greater}(x, y) = \begin{cases} \underline{\text{true}} \text{ if } x \text{ is greater than } y \\ \underline{\text{false}} \text{ if } x \text{ is equal to or smaller than } y \end{cases} \quad **$$

The value of a relation is often written with the help of an operator such as > or ≠; for instance, one generally writes

$$x > y$$

instead of: greater (x, y). **

In the sequel relations will be introduced, the arguments of which are ordered sequences rather than objects (Sections 3.4.3 and 3.4.4); these relations correspond to the generalized notion of function indicated in Section 1.4.6.

1.5.2. Non-negative integers

Let from now on $\underline{N}$ denote the set of all nni's.

1.5.2.1. $\underline{N}$-functions

A *natural function* (or: *$\underline{N}$-function*) with n arguments is any n-ary function the domain and range of which are subsets of $\underline{N}^n$ and $\underline{N}$ respectively.

Examples of $\underline{N}$-functions are the functions F_1, F_2 and F_3 of Section 1.4.3. Another example is:

$$\text{diff}: \{ (x, y) \mid x, y \in \underline{N}, x \geq y \} \rightarrow \underline{N} : \text{diff}(x, y) = x - y \quad **$$

The value of a 2-ary $\underline{N}$-function is often written with the help of an operator such as + or − ; for instance one writes generally x − y instead of diff (x, y).

1.5.2.2. Total and partial $\underline{N}$-functions.

An n-ary $\underline{N}$-function is *total* if its domain is $\underline{N}^n$. Examples of total $\underline{N}$-functions are F_1 and F_2 of Section 1.4.3; counter-examples are F_3 of Section 1.4.3 and diff of Section 1.5.2.1.

(*) $\underline{N}$ is the set of all nni's.

Sometimes one wants to explicitely state that a given function is not necessary total; he then says that the function is *partial*. Note that, strictly speaking, the label "partial" attached to "function" does not add any information.

1.5.2.3. Complement

The *complement* of a set $\underline{S} \subset \underline{N}^n$ ($n \geq 1$) is its complement with respect to $\underline{N}^n$, i.e. is $\underline{N}^n - \underline{S}$.

1.5.3. Strings

These objects are treated in the next chapter.

CHAPTER 2 : SETS AND FUNCTIONS OF STRINGS

Abstract: Some definitions concerning strings, sequences of strings and sets of strings are introduced and a few properties are indicated. Particular attention is paid to the enumeration of all strings or ordered sequences of strings over a given vocabulary.

2.1. Definitions

A *vocabulary* is a finite non-empty set; the elements of this set are called *symbols*. Examples of vocabularies:

$$\underline{V} = \{ a, b \} ;$$

$$\underline{V}' = \{ 0, 1, 2, 3, 4, 5, 6, 7, 8, 9 \} ;$$

$$\underline{V}'' = \{ 0, 1, +, = \} .$$ **

A *string (over a vocabulary* $\underline{V}$*)* is the object obtained by concatenating (i.e. writing one after the other) some symbols of $\underline{V}$ (*). Examples:

a, aa, aabab are three strings over $\underline{V}$

10+11=101 is a string over $\underline{V}''$ **

The *empty string* is the string which consists of no symbols; in the sequel it is denoted ε.

The *(free) monoid generated by a vocabulary* $\underline{V}$ is the set of all strings over $\underline{V}$, including the empty string. The free monoid generated by $\underline{V}$ is denoted $\underline{V}^*$ (**). Example: $\underline{V}^*$ contains the strings ε, a, b, aa, ab, ba, bb, aaa, ... **

A *substring* ψ *of a given string* ϕ over $\underline{V}$ is a fraction of ϕ or, more precisely, is a string for which there exist strings ω_1 and ω_2 in $\underline{V}^*$ such that

$$\phi = \omega_1 \psi \omega_2$$

(*) A string over $\underline{V}$ is viewed by some authors as an ordered sequence of elements of $\underline{V}$ in the notation of which the parentheses and commas are omitted.

(**) $\underline{V}^*$ together with the operation conc of Section 2.2 is a monoid in the sense of modern algebra; the elements of $\underline{V}$ are the generators of this monoid.

A substring is *proper* when $\omega_1\omega_2 \neq \varepsilon$, i.e. when not simultaneously $\omega_1 = \varepsilon$ and $\omega_2 = \varepsilon$. Example: aba is a proper substring of ababaa (take $\omega_1 = \varepsilon$, ω_2 = baa, or take ω_1 = ab , ω_2 = a) but a non-proper substring of aba. **

It is important not to confuse a symbol with an occurrence of this symbol in a string. For instance, there exists only one symbol b (in the vocabulary $\underline{V}$ = { a, b }) but there are two different occurrences of the symbol b in the string bab. A similar remark holds for a string and its occurrence as a substring in another string: there is only one string aba but there are two occurrences of aba in the string ababaa.

As a final remark it should be stressed that ε is a string (of zero symbols), not a symbol! In particular, ε should not be confused with the "blank symbol" which is used to separate the words in a typewritten text. By the way, the use of such a symbol is avoided in this course or, more precisely, when in Section 3.4.1 the "blank symbol" is introduced for mnemonic purposes, it is given a representation similar to that of the other symbols, viz. B.

2.2. String functions

A *$\underline{V}$-(string) function* with n arguments is a function the domain and range of which are subsets of $\underline{V}^{*n}$ and $\underline{V}^*$ respectively.

Examples of $\underline{V}$-string function are:

(i) conc: $\underline{V}^{*2} \rightarrow \underline{V}^*$: conc (x, y) = xy;
examples: conc (ab, ba) = abba, conc (a, ε) = a

(ii) head: $\underline{V}^* - \{\varepsilon\} \rightarrow \underline{V}^*$: head (x) = the leftmost symbol of x;
examples: head (abab) = a, head (ε) is undefined

(iii) delete: $\underline{V} \times \underline{V}^* \rightarrow \underline{V}^*$: delete (a,x) = the string x from which all occurrences of the symbol a are deleted;
examples: delete (a, abab) = bb, delete (a,b) = b

(iv) reverse: $\underline{V}^* \rightarrow \underline{V}^*$: reverse (x) = the string x the symbols of which are written in reverse order;
example: reverse (abb) = bba, reverse (ε) = ε. In the sequel the notation x^R will replace reverse (x); the letter R may then be considered as a unary operator.

(v) palindrome: $\underline{V}^* \rightarrow \underline{V}^*$: palindrome (x) = conc (x,x^R);
example: palindrome (ab) = abba **

An n-ary $\underline{V}$-function is *total* when its domain is $\underline{V}^{*n}$. A function which is not necessarily total is sometimes said to be *partial* in order to stress this fact.

2.3. Further notations and definitions

The following notation will be used in the sequel: if a is a symbol and n an nni, a^n represents the string of n symbols a. Example : $\{a^n b^n \mid n \geq 0\}$ is the set the elements of which are ε, ab, aabb, aaabbb, ... **

The *length* of a string (over a given vocabulary) is the number of its symbols. The length of a string, say the string x, is denoted by l(x). Examples: $l(ab) = 2$, $l(\varepsilon) = 0$ (*). **

The *complement* of a set $\underline{S} \subset \underline{V}^{*n}$ is its complement with respect to $\underline{V}^{*n}$ i.e. $\underline{V}^{*n} - \underline{S}$.

The *setproduct* of two sets of strings $\underline{A}$ and $\underline{B}$ is noted $\underline{A}.\underline{B}$ and is by definition:

$$\underline{A}.\underline{B} = \{ xy \mid x \in \underline{A} , y \in \underline{B} \}$$

It is important not to mix up the notions of setproduct and cartesian product; remember that the latter is a set of pairs rather than a set of strings:

$$\underline{A} \times \underline{B} = \{ (x, y) \mid x \in \underline{A} , y \in \underline{B} \}.$$

2.4. The interpretation of strings

2.4.1. The use of strings

When writing about objects (e.g. nni's, truth values, cows) one is obliged to choose a representation.

This representation may be two-dimensional (e.g. a drawing) but is often a string (i.e. a quasi-unidimensional representation).

In order to avoid ambiguity these strings must be such that in a given representation there corresponds with each string a single object.

(*) l may also be considered as a function from $\underline{V}^*$ onto $\underline{N}$.

As an example, in the decimal notation the string 12 (which is a string of $\{ 0, 1, 2, 3, 4, 5, 6, 7, 8, 9 \}^*$) represents a natural number, viz. twelve. In the binary notation the string 1101 (which is a string of $\{ 0, 1 \}^*$) represents the same number. As another example, the string goto L represents a jump in the programming language Algol-60. **

This correspondence between strings and the objects they represent may be viewed as a function from the set of strings onto the set of objects. It is interesting to note that this function is not necessarily injective: in the decimal notation, for instance, the strings 12 and 012 represent the same number.

2.4.2. The interpretation of strings

Considering the object a string stands for (i.e. the object a string represents) is said to interprete the string or, equivalently, to consider its semantics.

These Notes are concerned with the study of strings but their (possible) interpretation is never referred to. Hence the definitions and properties of (sets and functions of) strings which are given here and in the sequel are independent from their interpretation i.e. are valid whatever their interpretation may be.

2.5. Alphabetic order

An *alphabetic order (in a vocabulary $\underline{V}$)* is a function

$$A : \underline{V} \twoheadrightarrow \{ i \mid i \in \underline{N} , 1 \leq i \leq \text{card } (\underline{V}) \}$$

Note that A is finite and injective.

Less formally, providing a vocabulary $\underline{V}$ with an alphabetic order consists in listing its symbols in some arbitrary order and associating with them the numbers 1, 2, ..., card ($\underline{V}$) respectively.

Example: An alphabetic order in { a, b, c } is

$$A = \{ (a, 1) , (b, 2) , (c, 3) \};$$

another alphabetic order in the same vocabulary is

$$B = \{ (a, 2) , (b, 3) , (c, 1) \}$$ **

It is easily seen that the number of different alphabetic orders in a given vocabulary $\underline{V}$ is (card ($\underline{V}$))!

2.6. Enumeration of strings and n-tuples of strings

2.6.1. An enumeration of the elements of a monoid

Let $\underline{V}$ be a vocabulary provided with an alphabetic order A.

It is possible to enumerate the elements of $\underline{V}^*$ (i.e. to draw up a list of an arbitrary large number of elements of $\underline{V}^*$) according to the following two rules:

(i) a string of length k $(1 \leq k)$ is preceded by all strings of length smaller than k;

(ii) a string of length k $(1 \leq k)$, say $\beta_1\beta_2\ldots\beta_k (\beta_1, \beta_2, \ldots, \beta_k \in \underline{V})$, is preceded by all strings $\alpha_1\alpha_2\ldots\alpha_k (\alpha_1, \alpha_2, \ldots, \alpha_k \in \underline{V})$ of length k for which there exists h $(1 \leq h \leq k)$ such that

$$\alpha_i = \beta_i \quad \text{for all } i < h$$

and $$A(\alpha_h) < A(\beta_h)$$

Less formally, rule (i) requires that first all strings of length 0 are listed, then all strings of length 1, then all strings of length 2, etc. Rule (ii) requires that the strings of the same length are listed "alphabetically", i.e. in the same way as the words in a dictionary.

Example: For $\underline{V} = \{ a, b \}$ with $A = \{ (a, 1) , (b, 2)\}$ the strings of $\underline{V}^*$ are enumerated as follows:

$$\varepsilon, a, b, aa, ab, ba, bb, aaa, aab, aba, \ldots$$

Other example: For $\underline{V} = \{ a, b \}$ with $A = \{(b, 1), (a, 2)\}$ the enumeration is:

$$\varepsilon, b, a, bb, ba, ab, aa, bbb, bba, bab, \ldots$$ **

2.6.2. Generalization

The method for enumerating the elements of $\underline{V}^*$ will now be generalized for $\underline{V}^{*n}$, $n \geq 1$.

Again, suppose that the vocabulary $\underline{V}$ is provided with an alphabetic order A. Consider now a symbol C, $C \notin \underline{V}$ and provide the vocabulary $\underline{V} \cup \{C\}$ with the alphabetic order

$$A' = A \cup \{ (C, \text{card } (\underline{V}) + 1) \}$$

The enumeration of the elements of $\underline{V}^{*n}$ may now proceed as follows;

(i) the elements of the monoid $(\underline{V} \cup \{C\})^*$ are enumerated as indicated in Section 2.6.1;

(ii) the strings which do not contain exactly n-1 occurrences of the symbol C are dropped from the list;

(iii) each string of the list obtained sub (ii), say

$$x_1Cx_2C...Cx_n \qquad (x_1, x_2, ..., x_n \in \underline{V}^*)$$

is replaced by the n-tuple

$$(x_1, x_2, ..., x_n)$$

It should be clear that the resulting list constitutes an enumeration of the elements of $\underline{V}^{*n}$; moreover, for n = 1 this enumeration coincides with that of Section 2.6.1.

Example: Consider $\underline{V} = \{a,b\}$ provided with the alphabetic order $A = \{(a,1) , (b,2)\}$. The enumeration of the elements of $(\underline{V} \cup \{C\})^*$ is:

ε, a, b, C, aa, ab, aC, ba, bb, bC, Ca, Cb, CC, aaa, aab,
aaC, aba, abb, abC, aCa, aCb, ... ;

dropping the strings which do not contain exactly one occurrence of the symbol C yields:

C, aC, bC, Ca, Cb, aaC, abC, aCa, aCb, ... ;

hence the enumeration of the elements of $\underline{V}^{*2}$ is:

(ε, ε) , (a, ε) , (b, ε) , (ε, a) , (ε, b) , (aa, ε) ,
(ab, ε) , (a, a) , (a, b) , ...

This enumeration is illustrated by Figure 1. **

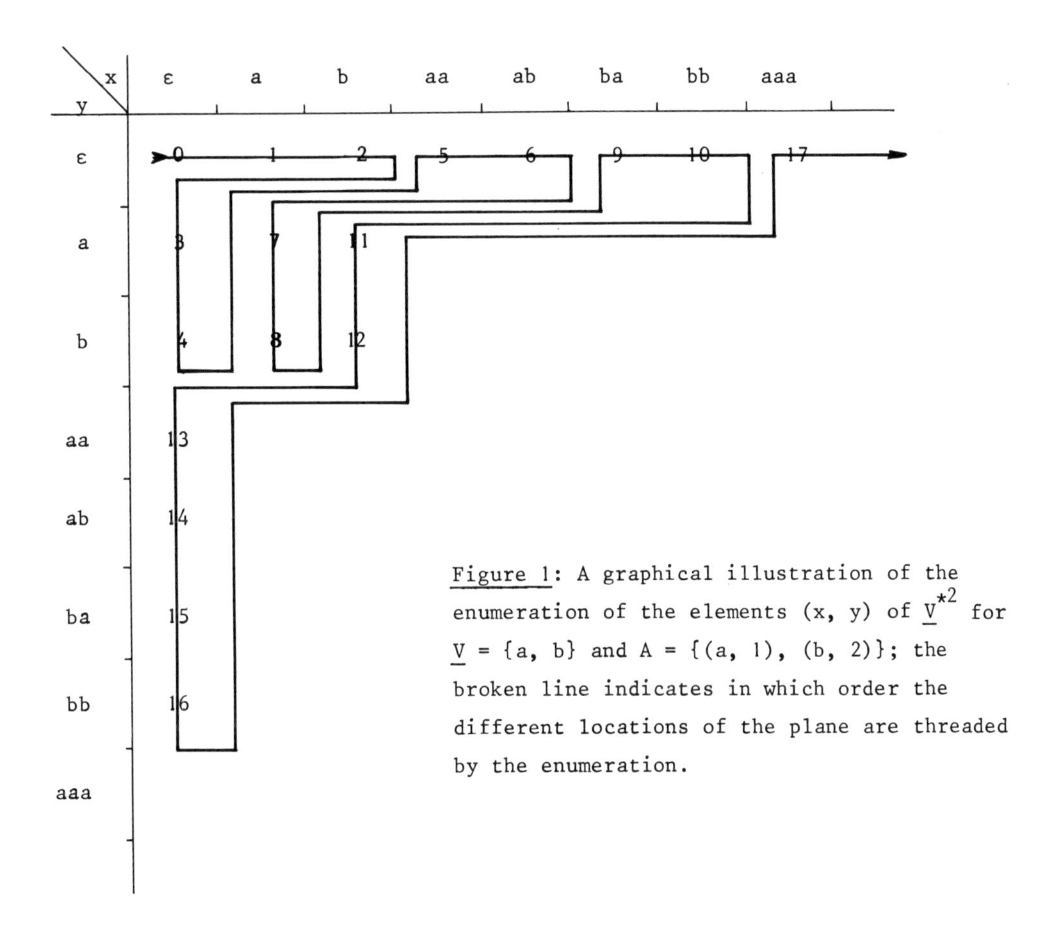

Figure 1: A graphical illustration of the enumeration of the elements (x, y) of $\underline{V}^{*2}$ for $\underline{V} = \{a, b\}$ and $A = \{(a, 1), (b, 2)\}$; the broken line indicates in which order the different locations of the plane are threaded by the enumeration.

2.7. Enumeration functions

Several classes of functions will now be introduced. They show how it is possible to relate in a one-to-one correspondence the non-negative integers, the strings over a vocabulary and the n-tuples of strings over a vocabulary. These functions are to be used in the subsequent Chapters.

2.7.1. The functions $\underline{V}^{*n}$-to-$\underline{N}$

Let $\underline{V}$ be a vocabulary provided with an alphabetic order A. Per definition, for $n \geqslant 1$:

$$(\underline{V}^{*n}, A)\text{-to-}\underline{N} : \underline{V}^{*n} \twoheadrightarrow \underline{N} :$$

$(\underline{V}^{*n}, A)\text{-to-}\underline{N}(x_1, x_2, \ldots, x_n) = k$, being understood that $(x_1, x_2, \ldots, x_n)$ is the $(k+1)^{\underline{th}}$ n-tuple in the enumeration of the elements of $\underline{V}^{*n}$.

Less formally, the function $(\underline{V}^{*n}, A)$-to-$\underline{N}$ associates with any n-tuple the number of its place in the enumeration described in Section 2.6.2.

When A is understood one may write

$$\underline{V}^{*n}\text{-to-}\underline{N}$$

instead of $(\underline{V}^{*n}, A)$-to-$\underline{N}$ and when n=1 one may write

$$\underline{V}^{*}\text{-to-}\underline{N}$$

instead of $\underline{V}^{*1}$-to-$\underline{N}$.

Examples: For $\underline{V} = \{ a, b \}$ and $A = \{ (a, 1) , (b, 2) \}$ one has, for instance (*):

$\underline{V}^{*}$-to-$\underline{N}$ $(\varepsilon) = 0$;

$\underline{V}^{*}$-to-$\underline{N}$ $(a) = 1$;

$\underline{V}^{*}$-to $\underline{N}$ $(b) = 2$;

$\underline{V}^{*}$-to-$\underline{N}$ $(aa) = 3$;

etc.

and:

$\underline{V}^{*2}$-to-$\underline{N}$ $(\varepsilon, \varepsilon) = 0$;

$\underline{V}^{*2}$-to-$\underline{N}$ $(a, \varepsilon) = 1$;

$\underline{V}^{*2}$-to-$\underline{N}$ $(b, \varepsilon) = 2$;

$\underline{V}^{*2}$-to-$\underline{N}$ $(\varepsilon, a) = 3$;

etc.

**

(*) cf. the examples of Section 2.6.1 and 2.6.2.

It is clear that $\underline{V}^{*n}$-to-$\underline{N}$ is a total injective function from $\underline{V}^{*n}$ onto $\underline{N}$. Hence, it makes sense to speak of the inverse functions

$$\overline{(\underline{V}^{*n}, A)\text{-to-}\underline{N}}\ ^{(j)} \qquad (1 \leq j \leq n)$$

One defines the function $\underline{N}$-to-($\underline{V}^{*}$, A) or, shortly, $\underline{N}$-to-$\underline{V}^{*}$ by:

$$\underline{N}\text{-to-}(\underline{V}^{*}, A) = \overline{(\underline{V}^{*}, A)\text{-to-}\underline{N}} \qquad ;$$

note that this function is a total injective function from $\underline{N}$ onto $\underline{V}^{*}$.

Example: For $\underline{V}$ and A defined as in the preceding example one has:

$\underline{N}$-to-$\underline{V}^{*}$ (0) = ε

$\underline{N}$-to-$\underline{V}^{*}$ (1) = a

$\underline{N}$-to-$\underline{V}^{*}$ (2) = b

$\underline{N}$ to-$\underline{V}^{*}$ (3) = aa

etc. **

One may consider the function $\underline{V}^{*}$-to-$\underline{N}$ as a generalization of the alphabetic order A (*); in fact:

$$A : \underline{V} \twoheadrightarrow \{ i \mid 1 \leq i \leq \text{card}(\underline{V}) \} : A(x) = \underline{V}^{*}\text{-to-}\underline{N}(x).$$

Further properties of the function $\underline{V}^{*n}$-to-$\underline{N}$ and, in particular, a procedure allowing to compute its value will be indicated in Section 2.8.

2.7.2. The functions $\underline{V}^{*n}$-to-$\underline{W}^{*}$

If $\underline{V}$ and $\underline{W}$ are vocabularies provided with the alphabetic orders A and B respectively, then one defines for $n \geq 1$:

$$(\underline{V}^{*n}, A)\text{-to-}(\underline{W}^{*}, B) : \underline{V}^{*n} \twoheadrightarrow \underline{W}^{*}$$

$$(\underline{V}^{*n}, A)\text{-to-}(\underline{W}^{*}, B)\ (x_1, x_2, \ldots, x_n) =$$

$$\underline{N}\text{-to-}(\underline{W}^{*}, B)\ ((\underline{V}^{*n}, A)\text{-to-}\underline{N}(x_1, x_2, \ldots, x_n))$$

Less formally, the function ($\underline{V}^{*n}$, A)-to-($\underline{W}^{*}$, B) associates with each n-tuple of $\underline{V}^{*n}$ the string of $\underline{W}^{*}$ which occurs at the same place in the enumeration.

(*) More precisely, $\underline{V}^{*}$-to-$\underline{N}$ is an extension of A, "extension" being used in the sense of modern algebra.

Again, when A and B are understood one may write

$$\underline{V}^{*n}\text{-to-}\underline{W}^{*}$$

instead of $(\underline{V}^{*n}, A)$-to-(W^{*}, B) and when n=1 one may write

$$\underline{V}^{*}\text{-to-}\underline{W}^{*}$$

instead of $\underline{V}^{*1}$-to-$\underline{W}^{*}$.

Examples:

(i) $\underline{V} = \underline{W} = \{ a, b \}$, $A = B = \{ (a, 1) , (b, 2) \}$, $n = 2$; then

$$\underline{V}^{*2}\text{-to-}\underline{V}^{*}\ (\varepsilon, \varepsilon) = \varepsilon$$

$$\underline{V}^{*2}\text{-to-}\underline{V}^{*}\ (a, \varepsilon) = a$$

$$\underline{V}^{*2}\text{-to-}\underline{V}^{*}\ (b, \varepsilon) = b$$

$$\underline{V}^{*2}\text{-to-}\underline{V}^{*}\ (\varepsilon, a) = aa$$

etc ;

this may be illustrated by the following table in which the elements of $\underline{V}^{*2}$, $\underline{N}$ and $\underline{V}^{*}$ are enumerated:

$\underline{V}^{*2}$	$\underline{N}$	$\underline{V}^{*}$
$(\varepsilon, \varepsilon)$	0	ε
(a, ε)	1	a
(b, ε)	2	b
(ε, a)	3	aa
(ε, b)	4	ab
⋮	⋮	⋮

(ii) $\underline{V} = \{ a, b \}$, $\underline{W} = \{ a \}$, $A = \{ (a, 1), (b, 2) \}$, $B = \{ (a, 1) \}$, $n = 1$; then:

$$\underline{V}^{*}\text{-to-}\underline{W}^{*}\ (\varepsilon) = \varepsilon$$

$$\underline{V}^{*}\text{-to-}\underline{W}^{*}\ (a) = a$$

$$\underline{V}^{*}\text{-to-}\underline{W}^{*}\ (b) = aa$$

$$\underline{V}^{*}\text{-to-}\underline{W}^{*}\ (aa) = aaa$$

etc.

or, in general, $\underline{V}^{*}$-to-$\underline{W}^{*}$ (x) is a string of $\underline{V}^{*}$-to-$\underline{N}$ (x) symbols a. **

It is clear that $\underline{V}^{*n}$-to-$\underline{W}^{*}$ is a total injective ($\underline{V} \cup \underline{W}$)-string function from $\underline{V}^{*n}$ onto $\underline{W}^{*}$. Hence it makes sense to speak of its inverse functions. Note in particular that

$$\overline{\underline{V}^{*}\text{-to-}\underline{W}^{*}} = \underline{W}^{*}\text{-to-}\underline{V}^{*}$$

The functions $\underline{V}^{*n}$-to-$\underline{W}^{*}$ do not only allow to replace (the study of) n-tuples by (that of) strings, but also to replace one vocabulary by another (viz. $\underline{V}$ by $\underline{W}$). How far-reaching the consequences of this are will appear in Chapters 4, 5 and 6; in particular, the functions $\underline{V}^{*n}$-to-$\underline{W}^{*}$ will play an essential role in the proof of the existence of non-computable functions (Section 5.1).

2.8. Calculating the value of the enumeration functions

The enumeration functions have been defined in Section 2.7 but the way in which their value for given arguments may be obtained has not been stated explicitly. A procedure will now be described which allows to calculate the value of the function ($\underline{V}^{*n}$, A)-to-$\underline{N}$ for an arbitrary given n-tuple of arguments ($\underline{V}$ and A are a vocabulary and an alphabetic order over it, $n \geq 1$). The case of the function ($\underline{V}^{*n}$, A)-to-($\underline{W}^{*}$, B) and of the inverse functions may be treated in a similar way.

The procedure makes use of two variables, viz. s and σ, to which respectively an nni and a string may be assigned as a value (*). The principle of its working consists in the enumeration of the n-tuples of $\underline{V}^{*n}$ according to the method indicated in Section 2.6.2 until the argument shows up. More precisely, when an n-tuple $(x_1, x_2, \ldots, x_n) \in \underline{V}^{*n}$ is presented to it, the procedure successively performs the following steps, C and A' being defined as in Section 2.6.2:

(i) put (**) $s = 0$ and $\sigma = C^{n-1}$;

(ii) if $\sigma = x_1Cx_2C\ldots Cx_n$, stop; the result is the value of s;

(iii) if $\sigma \in \{C\}^{*}$, put

$$\sigma = (\overline{A'}(1))^{1(\sigma)+1} \qquad (***)$$

and go to step (v);

(*) "Variable" is used here in the same sense as in a programming language.

(**) Put $s = 0$ means: assign the value 0 to the variable s.

(***) $\overline{A'}$ is the inverse function of the alphabetic order A', $1(\sigma)$ is the length of the (old) value of σ.

(iv) if $\sigma \notin \{C\}^*$, determine $d \in \underline{V}$, $y \in (\underline{V} \cup \{C\})^*$, $z \in \{C\}^*$ such that

$$\sigma = ydz \quad ;$$

put $\sigma = y\,\overline{A'}(A(d)+1)\,(\overline{A'}(1))^{l(z)}$;

(v) if σ does not contain exactly n-1 occurrences of the symbol C, go to step (iii);

(vi) if σ contains exactly n-1 occurrences of the symbol C, increase s by 1(*) and go to step (ii).

The working of this procedure may be illustrated by the following comments: step (i) assigns initial values to the variables; step (ii) checks whether the given n-tuple $(x_1, x_2, \ldots, x_n)$ is reached and, if so, makes the procedure come to an end; step (iii) increases the length of the string by one symbol according to rule (i) of Section 2.6.1; step (iv) determines the alphabetically next string according to rule (ii) of Section 2.6.1; step (v) corresponds with step (ii) of Section 2.6.2.

It is clear that for any $(x_1, x_2, \ldots, x_n) \in \underline{V}^{*n}$ this procedure comes to an end and that it delivers the value

$$(\underline{V}^{*n}, A)\text{-to-}\underline{N}\ (x_1, x_2, \ldots, x_n)$$

as a result.

To a certain extent the procedure is of pure theoretical interest because for non-trivial arguments its execution time increases beyond reasonable limits. It is then necessary to replace the procedure by a more sophisticated one which avoids that, at each calculation, one has to work through all "preceding" n-tuples. The description of such procedures is beyond the scope of this course: it may suffice to indicate that for n=1 and $a_1, a_2, \ldots, a_k \in \underline{V}$, $k \geq 0$:

$$(\underline{V}^*, A)\text{-to-}\underline{N}\ (a_1 a_2 \ldots a_k) = \sum_{i=1}^{k} A(a_i) \times (\text{card}\ (\underline{V}))^{k-i} \quad (**)$$

and that, consequently, the procedure reduces to the calculation of the right member of this equality.

(*) i.e. put s = s+1.

(**) This formula shows that $(\underline{V}^*, A)$-to-$\underline{N}$ is similar to (but not identical with) the interpretation of a string as a number in base card $(\underline{N})$; by the way, if it were identical, $(\underline{V}^*, A)$-to-$\underline{N}$ would not be injective (see Section 2.4.1).

CHAPTER 3 : COMPUTABLE FUNCTIONS

Abstract: The Turing machine is introduced both by describing its physical model and by giving a formal (algebraic) definition for it. The notion of computability of ($\underline{V}$-string) functions is derived from it. Finally, it is proved that any Turing machine may be replaced by a Turing machine of a particular type, called "normal" Turing machine.

3.1. Historical background

An axiomatic theory such as geometry essentially consists of a set of axioms and a set of deduction rules. From these axioms one may obtain "automatically" (i.e. by "blindly" applying the deduction rules) all theorems of the theory, e.g. all theorems of geometry or, in other words, all true statements concerning geometry.

For a long time and, in particular, during the nineteenth century mathematicians have tried to set up axiomatic theories for branches of mathematics other than geometry. This was tried in particular for the theory of natural numbers - which, by the way, contains the predicate calculus as a constituent part.

In 1931 Gödel proved that the theory of natural numbers cannot be defined by an axiomatic theory. More precisely, he proved that an axiomatic theory for natural numbers would be either incomplete (i.e. not all true statements are obtained in the form of a theorem) or inconsistent (i.e. the theory leads to theorems expressing contradictory statements). He thus showed that there exist sets (viz. the sets of all true statements concerning natural numbers) which are in some sense not enumerable or, equivalently, that there exist functions (viz. the functions which map the true statements concerning natural numbers into the value true and the false statements concerning natural numbers into the value false) which are in some sense not computable.

About 1936 Turing, Church and Post gave independently from each other an apparently different definition for the notion of computability; more precisely, Turing introduced the Turing machine, Church drew up the recursive function theory and Post introduced a symbol manipulation system, now called Post system. Actually, a few years later these three definitions were shown to be mathematically equivalent.

In the two last decades the study of computability developed in two directions. The first direction tries to gain deeper insight into the notion of computability and is connected with the study of the foundations of mathematics.

The second direction is concerned with the study of notions which are less general than computability such as finite state automata (being a special type of Turing machines) or context-free grammars (being a special type of Post systems), and is connected with mathematical linguistics and computer programming.

3.2. The basic idea of Turing

In Section 1.4.1 a function has been defined in a "static" way: the definition does not (explicitly) give a "rule" (or "constructive procedure", or: "algorithm", or: "program") for calculating the value of the function. But, as already indicated in Section 1.4.4, the definition of a function is rather useless as long as one does not possess such a rule, i.e. as long as one is not given a "constructive definition" of the function; this is in particular true when the domain of the function is not finite. Of course, such a rule must be precise; actually, it has to be so precise that nothing is left to the initiative of the (human) computer or, equivalently, that a clerk or a machine may execute it.

The idea of Turing is to define a function with the help of a machine. Each of these machines is defined by a set of "instructions"; the working of the machine consists in an "application" of these instructions and is interpreted as being the calculation of the value of the function for given arguments.

Such a machine, called Turing machine, is said to be a "mathematical" machine: this means that it might be constructed but that no implementation problems (such as cost, maintenance, etc.) nor efficiency problems (such as speed) are considered.

A Turing machine is essentially a string manipulating machine, i.e. for a given vocabulary $\underline{V}$ it defines a $\underline{V}$-string function. These strings may of course be given an interpretation (e.g. nni's or truth values) and hence Turing machines may be used for the definition of $\underline{N}$-functions, of predicates, etc. Nevertheless, as already indicated strings will not be interpreted in this course and hence Turing machines will exclusively be used for the definition of string functions.

3.3. Physical model

The structure and the working of a Turing machine will now be described with the help of a physical model. The actual definition of a Turing machine will be given in Section 3.4 but the (formal) definition given there is rather difficult to grasp when not having the (informal) physical-model definition in mind.

3.3.1. The machine

A Turing machine essentially consists of a "control unit", a "tape" and a "read-write device" (Figure 2).

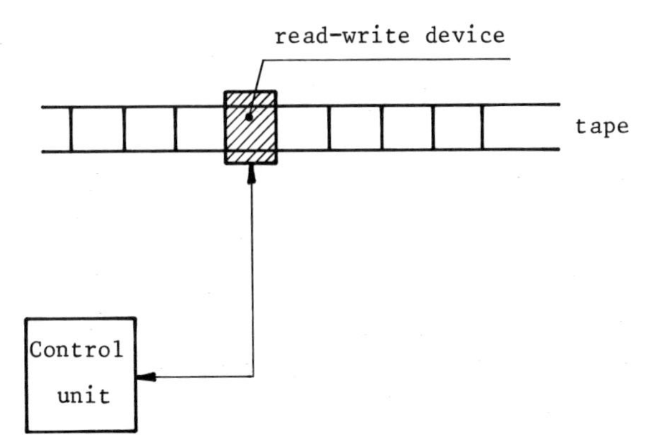

Figure 2: The physical model of a Turing machine

The control unit which "makes the machine run" is at each moment in a certain "state" out of a finite number of possible states; moreover it "contains" a set of "instructions" which at each moment univocally determine the working of the Turing machine; by the way, different Turing machines differ from each other essentially by the nature of these instructions.

The tape is divided into squares; each square is able to contain a symbol.

The read-write device covers exactly one square of the tape. It can read the symbol contained by this square; it can also write a symbol on this square while erasing (by "overwriting") the symbol contained previously by it; finally the read-write device can move over one square to the left or to the right, the tape itself being immovable.

3.3.2. Working

A Turing machine is started with the control unit in a particular state called the "start state" and with a string on its tape.

The working consists in the iterative execution of an elementary step. Each such step consists in:

(i) the reading of a symbol, viz. the symbol under the head of the read-write device;

(ii) the determination of the applicable instruction; this is done on the base of the state of the control unit and of the result of the read operation;

(iii) the execution of the applicable instruction which consists in performing successively the three following operations:

  (a) putting the control unit into a new state;

  (b) writing a symbol on the tape, - thus overwriting the symbol read during step (i);

  (c) moving the read-write device over one square to the left, moving it over one square to the right or keeping it fixed.

The Turing machine halts when no instruction is applicable.

3.3.3. The definition of a function

A Turing machine defines a (string) function thanks to the following interpretation of its working.

The string on the tape at the start of the Turing machine is interpreted as (a particular value for) the arguments of the function. The string on the tape after halting is interpreted as the value of the function for these arguments. The fact that the machine does not halt but continues working forever is interpreted as the value of the function being undefined for the given arguments.

A Turing machine thus constitutes a "rule" which (allows to) calculate the value of a function for given arguments, i.e. it constitutes a "constructive definition" of a (partial) function.

3.3.4. The memory of a Turing machine

One may consider a Turing machine as being constituted by two parts: a control unit which makes the machine run in a step-by-step way, and a memory (*). The memory is itself constituted by the (current) state of the control unit and by the contents of the tape. That the state constitutes a memory springs from the fact that each state depends (among others) on the preceding state and hence tells "something" about the history of the working. That the tape contents constitutes

---

(*) Note the similarity with an electronic computer.

a memory is clear: it is possible to write a symbol on the tape and - after some moves in one direction followed by some moves in the other direction - read it at some later moment.

The set of states may be considered to provide a finite memory because there are only a finite number of different possible states; in other words the control unit can only distinguish between a finite number of classes of "histories".

On the other hand the tape is said to provide an infinite memory because it is possible to write symbols on an arbitrary large number of squares; in fact, the Turing machine is allowed to run off the initially written part of the tape.

Actually, the tape of a Turing machine has not to be itself of infinite length but it should be possible to add a new square to it whenever the machine asks for it or, more precisely, whenever the read-write device moves off the (existing part of) the tape. The Turing machine may therefore be seen as a device provided with a finite tape together with a paper tape factory rather than as a device provided with an infinite tape; hence it is more precise to consider the tape as a memory of unbounded rather than of infinite dimensions. This fact is also illustrated by the following consideration: at each moment the used fraction of the tape (or, equivalently, the string contained by the tape) is of finite length.

3.4. Formal definition of a Turing machine

The Turing machine has been defined above informally with the help of a physical model. It will now be defined formally in a algebraic way.

In this formal definition references will be made to the physical model. It should be clear that these references are not part of the definition but are mere comments intended to facilitate the understanding.

3.4.1. The machine

A *($\underline{V}$-) Turing machine* (abbreviated: ($\underline{V}$-)TM) is defined by a 5-tuple ($\underline{V}$, $\underline{Q}$, $\underline{I}$, B, q_s) in which:

(i) $\underline{V}$ is a vocabulary;

(ii) $\underline{Q}$ is a finite set of symbols;

(iii) B is a symbol, $B \notin \underline{V}$;

(iv) q_s is an element of $\underline{Q}$;

(v) $\underline{I}$ is a finite set of 5-tuples (q, a, q', a', δ) with

- $q, q' \in \underline{Q}$
- $a, a' \in \underline{V} \cup \{B\}$
- $\delta \in \{L, R, O\}$ where L, R, O are three different arbitrary symbols

and with the following restriction:

- for any two different elements $(q_1, a_1, q'_1, a'_1, \delta_1)$ and $(q_2, a_2, q'_2, a'_2, \delta_2)$ of $\underline{I}$ it is not the case that simultaneously $q_1 = q_2$ and $a_1 = a_2$. (R)

$\underline{V}$ is called the *vocabulary of the Turing machine*, the elements of $\underline{Q}$ are called the *states*, B is called the *blank symbol*, q_s is called the *start state* and the elements of $\underline{I}$ are called the *instructions*. In the sequel an instruction is noted

$$(q, a) \rightarrow (q', a', \delta)$$

rather than

$$(q, a, q', a', \delta).$$

Interpreting these definitions on the physical model leads to the following considerations. $\underline{V}$ is the vocabulary used by the arguments and the value of the function to be defined; in other words, a $\underline{V}$-Turing machine is intended to define a $\underline{V}$-string function. $\underline{Q}$ is the set of states of the control unit. B is a particular symbol which is supposed to be contained by each new square. q_s is the state at the start of the Turing machine. Finally, an element

$$(q, a) \rightarrow (q', a', \delta)$$

means: when the control unit is in state q and the symbol read is a, then the control unit has to be put into state q', the symbol a has to be overwritten by the symbol a' and the read-write device has to move one square to the left, to move one square to the right or not to move, according to δ being L, R or O. The restriction (R) warrants that for a given state and a given read result at most one instruction is applicable, i.e. that at each moment the next step of the Turing machine is univocally defined.

3.4.2. Instantaneous description

3.4.2.1. Definition

An *instantaneous description* (abbreviated: ID) of a given Turing machine ($\underline{V}$, $\underline{Q}$, $\underline{I}$, B, q_s) is a triple (q, ϕ, ψ) with

- q is in $\underline{Q}$
- ϕ is in $(\underline{V} \cup \{B\})^*$
- ψ is in $(\underline{V} \cup \{B\})^* - \{\varepsilon\}$ (*)

Interpreted on the physical model an ID may be considered as a snapshot of the Turing machine: q is the state of the control unit, ϕ is the left-hand part of the used fraction of the tape, ψ is its right-hand part being understood that it encompasses the symbol under the read-write device (Figure 3); the condition $\psi \neq \varepsilon$ warrants that there is always a symbol under the read-write device.

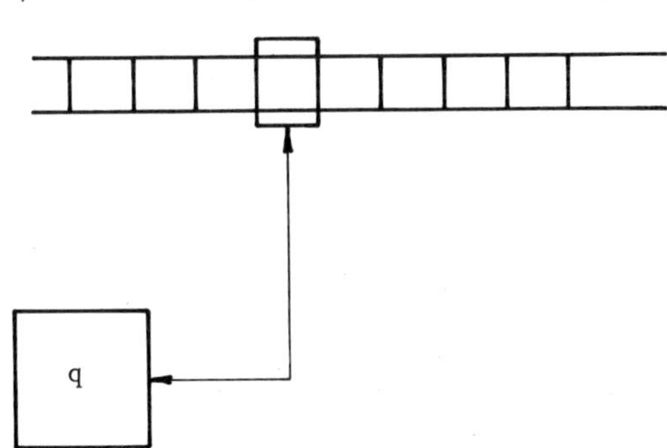

Figure 3: The instantaneous description of the Turing machine is (q, ϕ, ψ).

(*) Though an ID may have the same form as the right member of an instruction it has an essentially different meaning!

3.4.2.2. Applicable instruction

Let $\alpha = (q_o, \phi_o, \psi_o)$ be an arbitrary ID of a Turing machine $(\underline{V}, \underline{Q}, \underline{I}, B, q_s)$. The instruction

$$(q, a) \rightarrow (q', a', \delta)$$

of $\underline{I}$ is said to be *applicable* to the ID α if

$$q = q_o$$

and $$a = \text{head } (\psi_o) \qquad (*)$$

Note that the restriction (R) of Section 3.4.1 implies that at most one instruction of $\underline{I}$ is applicable to a given ID.

3.4.2.3. Initial and final instantaneous description.

An ID (q, ϕ, ψ) is *initial* if $q = q_s$ and $\phi = \varepsilon$.

An ID is *final* if no instruction is applicable to it.

3.4.3. The relation $\Longrightarrow$

3.4.3.1. The follower of an ID.

A construction will now be indicated which associates with a given arbitrary non-final ID (of a given arbitrary Turing machine) another ID called its follower.

Let $T = (\underline{V}, \underline{Q}, \underline{I}, B, q_s)$ be such a Turing machine and $\alpha = (q, \phi, a\psi)$ with $q \in \underline{Q}$, $a \in \underline{V} \cup \{B\}$, $\phi, \psi \in (\underline{V} \cup \{B\})^*$ such a non-final ID. Let further

$$(q, a) \rightarrow (q', a', \delta)$$

be the applicable instruction of α.

The *follower* of the ID α is then per definition the ID β obtained as follows:

(i) if $\delta = 0$ then $\beta = (q', \phi, a'\psi)$;

(ii) if $\delta = R$ then

(a) if $\psi \neq \varepsilon$, $\beta = (q', \phi a', \psi)$

(b) if $\psi = \varepsilon$, $\beta = (q', \phi a', B)$;

(*) The function head has been introduced in Section 2.2; note that the value head (ψ_o) is always defined as $\psi_o \neq \varepsilon$.

(iii) if $\delta = L$ then

(a) if $\phi \neq \varepsilon$, put $\phi = \phi_o c$ with $c \in \underline{V} \cup \{B\}$ and $\phi_o \in (\underline{V} \cup \{B\})^*$;

then $\beta = (q', \phi_o, ca'\psi)$;

(b) if $\phi = \varepsilon$, $\beta = (q', \varepsilon, Ba'\psi)$.

3.4.3.2. Interpretation on the physical model

Constructing the follower of an ID consists in executing an elementary step.

The case $\delta = 0$ is illustrated by the Figure 4a; the cases $\delta = R$, $\psi \neq \varepsilon$ and $\delta = R$, $\psi = \varepsilon$ are illustrated respectively by the Figures 4b and 4c.

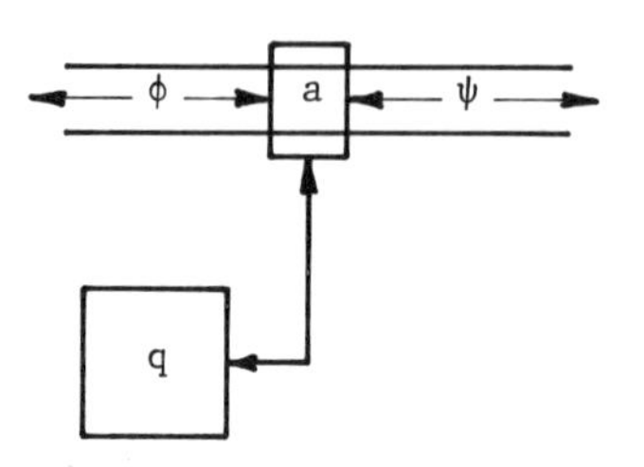

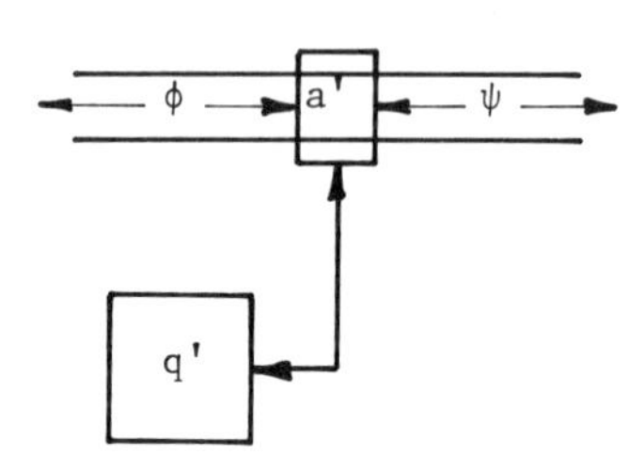

Figure 4a: $\delta = 0$

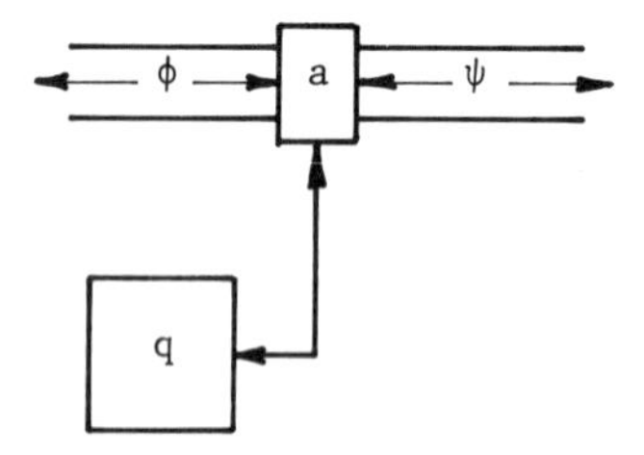

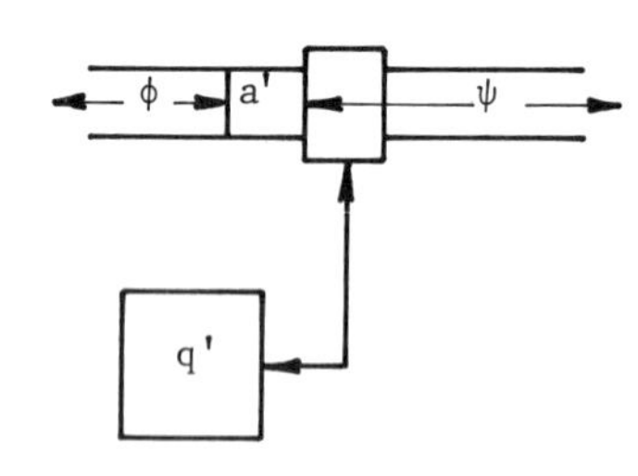

Figure 4b: $\delta = R$, $\psi \neq \varepsilon$

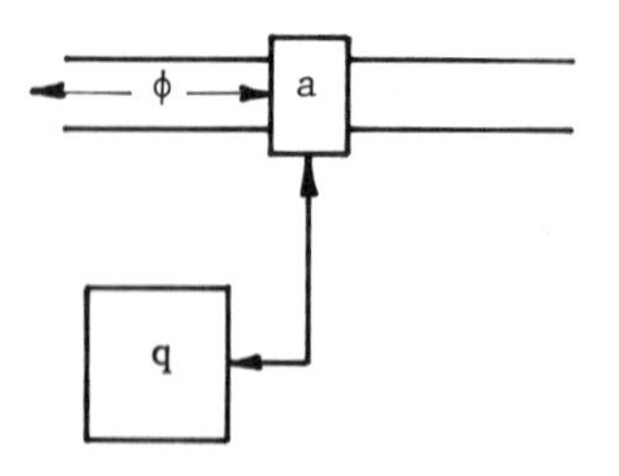

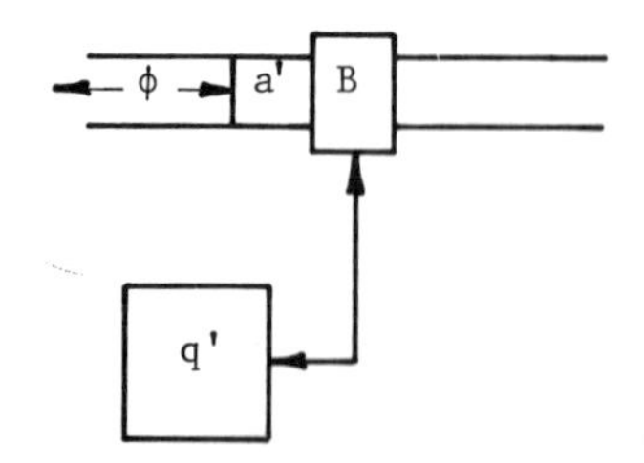

Figure 4c: $\delta = R$, $\psi = \varepsilon$

3.4.3.3. The relation $\Longrightarrow$

The fact that the ID β is the follower of the ID α is written:

$$\alpha \underset{T}{\Longrightarrow} \beta$$

or, if the Turing machine T is understood:

$$\alpha \Longrightarrow \beta$$

One may consider $\Longrightarrow$ as a relation between ID's. This relation is of a rather particular nature because for a given ID α there exists at most one ID β such that $\alpha \Longrightarrow \beta$.

3.4.4. The relation $\overset{*}{\Longrightarrow}$

Per definition the relation

$$\alpha \underset{T}{\overset{*}{\Longrightarrow}} \beta$$

between the ID's α and β holds if either $\alpha = \beta$ or there exists a finite sequence of ID's

$$\gamma_1 , \gamma_2 , \ldots, \gamma_n \qquad (n > 1)$$

such that

$$\gamma_1 = \alpha$$

$$\gamma_n = \beta$$

and

$$\gamma_i \underset{T}{\Longrightarrow} \gamma_{i+1} \quad \text{for } 1 \leqslant i \leqslant n - 1 \qquad (*)$$

If T is understood one may also write

$$\alpha \overset{*}{\Longrightarrow} \beta$$

instead of $\alpha \underset{T}{\overset{*}{\Longrightarrow}} \beta$.

Interpreted on the physical model the relation $\overset{*}{\Longrightarrow}$ represents the consecutive execution of a certain number (possibly zero) elementary steps.

(*) The relation $\overset{*}{\Longrightarrow}$ constitutes the reflexive transitive closure of the relation $\Longrightarrow$.

3.4.5. The functions defined

The *n-ary function* $f_{T,n}$ *defined by the* $\underline{V}$*-Turing machine* $T = (\underline{V}, \underline{Q}, \underline{I}, B, q_s)$ is the n-ary $\underline{V}$-string function

$f_{T,n} = \{ (x_1, x_2, \ldots, x_n, \text{delete}(B, yz)) \mid$ there exists
$q \in \underline{Q}$ such that $(q_s, \varepsilon, x_1Bx_2B\ldots Bx_nB) \overset{*}{\underset{T}{\Longrightarrow}} (q,y,z)$, and (q,y,z) is a final ID $\}$. (*)

Less formally, if there exists a state q such that (q,y,z) is final then

$$f_{T,n}(x_1, x_2, \ldots, x_n) = \text{delete}(B, yz) \qquad ,$$

otherwise

$$f_{T,n}(x_1, x_2, \ldots, x_n) \text{ is undefined.}$$

The interpretation on the physical model is as follows. The Turing machine is started with the arguments on its tape, the arguments being separated one from the other by the symbol B. When the Turing machine halts, the contents of the tape from which the occurrences of the symbol B are deleted is the value of the function for the given arguments; if the Turing machine does not halt, the value of the function for the given arguments is undefined.

A Turing machine T defines an infinity of functions, namely $f_{T,1}$, $f_{T,2}$, $f_{T,3}$, ... It should be clear that a particular Turing machine is generally designed in order to define one of these functions; the other functions are then not considered and are, moreover, often trivial.

One may wonder why in the definition of $f_{T,n}$ the last argument x_n in the initial ID $(q_s, \varepsilon, x_1B\ldots Bx_nB)$ is followed by the symbol B. The reason is the following: for the case $n = 1$, $x_1 = \varepsilon$, the triple $(q_s, \varepsilon, x_1B\ldots Bx_n)$ is not an ID because its third element is ε.

(*) The function delete is defined in Section 2.2.

3.4.6. The constructive nature of the function definition

That a Turing machine $T = (\underline{V}, \underline{Q}, \underline{I}, B, q_s)$ provides a constructive definition (in the sense of Section 3.2) for the function $f_{T,n}$ clearly appears from the comments which accompany the preceding sections and which relate the formal definitions to the physical model.

The constructive nature of the definition of the function $f_{T,n}$ also appears from the following procedure which "simulates" the working of the Turing machine; this procedure constitutes a constructive definition of $f_{T,n}$ by making correspond with an arbitrary n-tuple $(x_1, x_2, \ldots, x_n) \in \underline{V}^{*n}$ the value $f_{T,n}(x_1, x_2, \ldots, x_n)$ when it is defined, and by not coming to an end when it is undefined:

(i) put $\alpha = (q_s, \varepsilon, x_1Bx_2B\ldots Bx_nB)$; (*)

(ii) determine by considering successively the different elements of $\underline{I}$ whether or not α has a follower;

(iii) if α has no follower, stop; the result is delete (B, yz), y and z being defined by $\alpha = (q, y, z)$;

(iv) if α has a follower, construct it and call it β;

(v) put $\alpha = \beta$ and go to step (ii).

3.5. Examples of Turing machines

3.5.1. First example

Consider $T = (\underline{V}, \underline{Q}, \underline{I}, B, q_s)$ with

$\underline{V} = \{a, b\}$

$\underline{Q} = \{q_o, q, q_f\}$

$B = \$$ (**)

$q_s = q_o$ (**)

(*) α is a variable to which a value may be assigned (cf. a variable in a programming language).

(**) This equality expresses that the value of the variable B (resp. q_s) is the symbol \$ (resp. q_o).

I consists of the following six instructions:

$(q_o, a) \rightarrow (q, a, R)$ (1)

$(q_o, b) \rightarrow (q, b, R)$ (2)

$(q_o, \$) \rightarrow (q_o, \$, 0)$ (3)

$(q, a) \rightarrow (q, \$, R)$ (4)

$(q, b) \rightarrow (q, \$, R)$ (5)

$(q, \$) \rightarrow (q_f, \$, 0)$ (6)

It is now claimed that $f_{T,1}$ is the {a,b}-string function head defined in Section 2.2.(ii). This statement which is not proved here may be illustrated by the following two examples:

(i) $f_{T,1}(aba) = a$ because

$(q_o, \varepsilon, aba\$)$

$\Longrightarrow (q, a, ba\$)$ by application of the instruction (1)

$\Longrightarrow (q, a\$, a\$)$ by application of the instruction (5)

$\Longrightarrow (q, a\$\$, \$)$ by application of the instruction (4)

$\Longrightarrow (q_f, a\$\$, \$)$ by application of the instruction (6)

(ii) $f_{T,1}(\varepsilon)$ is undefined because

$(q_o, \varepsilon, \$)$

$\Longrightarrow (q_o, \varepsilon, \$)$ by application of the instruction (3)

$\Longrightarrow (q_o, \varepsilon, \$)$ by application of the instruction (3)

etc.

From these two examples the role of the different instructions of I appears to be as follows:

- the instructions (1) and (2) recognize the leftmost symbol;
- the instruction (3) leads into an "infinite loop" for the argument ε;
- the instructions (4) and (5) erase the "tail" of the string;
- the instruction (6) leads to a final ID.

Note that replacing the instruction (3) by, for instance, the instruction

$$(q_o, \$) \rightarrow (q_o, a, R)$$

leads to a Turing machine defining the same 1-ary function: for the argument ε it never halts while filling the right fraction of its tape with (an infinite number of occurrences of) the symbol a.

3.5.2. Second example

Consider the function string-subtraction over the vocabulary $\underline{V} = \{1\}$ defined as follows:

$$\text{ssub}: \{ (1^p, 1^q) \mid p \geq q \geq 0 \} \rightarrow \{1\}^* : \text{ssub}(1^x, 1^y) = 1^{x-y}.$$

The 2-ary function defined by the Turing machine $T = (\underline{V}, \underline{Q}, \underline{I}, B, q_s)$ described below is claimed to be ssub:

$\underline{V} = \{1\}$

$\underline{Q} = \{q_s, q_1, q_2, q_3, q_4, q_5, q_6\}$

$B = B$ (*)

$q_s = q_s$ (*)

$\underline{I}$ consists of the following 12 instructions the role of which is indicated to the right:

$(q_s, 1) \rightarrow (q_s, 1, R)$	locate the separating B
$(q_s, B) \rightarrow (q_1, B, R)$	
$(q_1, 1) \rightarrow (q_1, 1, R)$	locate the rightmost 1
$(q_1, B) \rightarrow (q_2, B, L)$	
$(q_2, 1) \rightarrow (q_3, B, L)$	erase the rightmost 1
$(q_2, B) \rightarrow (q_4, B, 0)$	stop

(*) This equality expresses that the value of the variable B (resp. q_s) is the symbol B (resp. q_s). As a difference with the example of Section 3.5.1 (see the footnote of this Section), B (resp. q_s) is now used for denoting two different things: a variable and a value of a variable. This ambiguity in the notation is not harmful provided the equality $B = B$ (resp. $q_s = q_s$) is interpreted correctly.

$(q_3, 1) \to (q_3, 1, L)$ $(q_3, B) \to (q_5, B, L)$	locate the separating B
$(q_5, 1) \to (q_5, 1, L)$ $(q_5, B) \to (q_6, B, R)$	locate the leftmost 1
$(q_6, 1) \to (q_s, B, R)$	erase the leftmost 1
$(q_6, B) \to (q_6, B, 0)$	loop

The working of this Turing machine is illustrated by the three following examples:

(i) ssub (11, 1) = 1 because

$$(q_s, \varepsilon, 11B1B)$$
$$\Longrightarrow (q_s, 1, 1B1B)$$
$$\Longrightarrow (q_s, 11, B1B)$$
$$\Longrightarrow (q_1, 11B, 1B)$$
$$\Longrightarrow (q_1, 11B1, B)$$
$$\Longrightarrow (q_2, 11B, 1B)$$
$$\Longrightarrow (q_3, 11, BBB)$$
$$\Longrightarrow (q_5, 1, 1BBB)$$
$$\Longrightarrow (q_5, \varepsilon, 11BBB)$$
$$\Longrightarrow (q_5, \varepsilon, B11BBB)$$
$$\Longrightarrow (q_6, B, 11BBB)$$
$$\Longrightarrow (q_s, BB, 1BBB)$$
$$\Longrightarrow (q_s, BB1, BBB)$$
$$\Longrightarrow (q_1, BB1B, BB)$$
$$\Longrightarrow (q_2, BB1, BBB)$$
$$\Longrightarrow (q_4, BB1, BBB)$$

(ii) ssub (ε, ε) = ε because

$$(q_s, \varepsilon, BB)$$
$$\Longrightarrow (q_1, B, B)$$
$$\Longrightarrow (q_2, \varepsilon, BB)$$
$$\Longrightarrow (q_4, \varepsilon, BB)$$

(iii) ssub (ε, 1) is undefined because

$$(q_s, \varepsilon, B1B)$$
$$\Longrightarrow (q_1, B, 1B)$$
$$\Longrightarrow (q_1, B1, B)$$
$$\Longrightarrow (q_2, B, 1B)$$
$$\Longrightarrow (q_3, \varepsilon, BBB)$$
$$\Longrightarrow (q_5, \varepsilon, BBBB)$$
$$\Longrightarrow (q_6, B, BBB)$$
$$\Longrightarrow (q_6, B, BBB)$$

etc.

3.6. Computable functions

Per definition, a (partial) n-ary $\underline{V}$-string function F is *(partial) computable* if there exists a $\underline{V}$-Turing machine T such that $f_{T,n} = F$.

Less formally, a function is computable if there exists a Turing machine defining it.

A *total computable* function is defined similarly.

3.7. The thesis of Turing

Turing emitted the following thesis: each function for which there exists a "rule" (or: "constructive procedure") to compute its value is a computable function (*). Note that this thesis cannot be proved because it connects a mathematical notion (viz. the Turing machine) to an empirical notion (viz. the "rule"). Hence it bears similarities with a thesis (or: "law") asserting that a certain mathematical equation is applicable to a physical phenomenon.

Apart from intuition the following facts corroborate the validity of the thesis of Turing: first, no counterexamples have yet been found; second, other definitions of computable functions resulting from completely different approaches (such as the recursive function theory, the Post systems or the Markov algorithms)

(*) The notion "rule" has been introduced in Sections 1.4.4 and 3.2. The converse of Turing's thesis was discussed in Section 3.4.6.

have all been shown to be mathematically equivalent with the definition of computable function based on Turing machines (*).

Showing that a function is computable is done by constructing a Turing machine for it. Actually, the construction of a Turing machine is generally a lengthy task and, moreover, the proof that it actually defines the function intended is tedious. In the sequel the proof of computability will therefore often be reduced to the informal description of a procedure for which it is intuitively clear that it is constructive; the function defined by this constructive procedure is then computable by virtue of the thesis of Turing.

For instance, the enumeration functions of Section 2.7 (and their inverse functions) can easily be defined by such constructive procedures; this is considered to be a sufficient argument for showing that these functions are or, more precisely, may be proved to be computable.

3.8. Normal Turing machines

3.8.1. Definition

Let $n \geq 1$ be an integer, $T = (\underline{V}, \underline{Q}, \underline{I}, B, q_s)$ a $\underline{V}$-Turing machine.

Per definition, the Turing machine T is *(n-)normal* if there exists a state $q_f \in \underline{Q}$ which satisfies the following condition:

for each ID, say (q, ϕ, ψ), for which there exist $x_1, x_2, \ldots, x_n \in \underline{V}^*$ such that

$$(q_s, \varepsilon, x_1Bx_2B\ldots Bx_nB) \overset{*}{\Longrightarrow} (q, \phi, \psi)$$

it is the case that:

(i) $q \neq q_f$ when (q, ϕ, ψ) is not a final ID;

(ii) $q = q_f$, $\phi \in \{B\}^*$ and $\psi \in \underline{V}^*.\{B\}^*$ when (q, ϕ, ψ) is a final ID (**).

Less formally, an n-normal Turing machine calculating the value of an n-ary function halts if and only if it is in a particular state; moreover, the value of the function defined is not intermixed with blank symbols and its leftmost symbol is under the head of the read-write device.

(*) This equivalence between Turing machines and, for instance, Markov algorithms has to be understood as follows: with each Turing machine it is possible to associate a Markov algorithm defining the same n-ary function, and vice versa.

(**) $\underline{V}^*.\{B\}^*$ denotes the set product of $\underline{V}^*$ and $\{B\}^*$ (see Section 2.3).

3.8.2. Lemma 3.1: For any $\underline{V}$-TM T one may construct a $\underline{V}$-TM S which defines the same $\underline{V}$-string functions and which contains a particular state p such that an ID of S is final if and only if the state of this ID is p.

Proof.

If $T = (\underline{V}, \underline{Q}, \underline{I}, B, q_s)$ is the given $\underline{V}$-TM, construct the $\underline{V}$-TM $S = (\underline{V}, \underline{Q}_s, \underline{I}_s, B, q_s)$ with:

$$\underline{Q}_s = \underline{Q} \cup \{p\}, \ p \notin \underline{Q}$$

$$\underline{I}_s = \underline{I} \cup \{(q, a) \to (p, a, 0) \mid q \in \underline{Q}, a \in \underline{V} \cup \{B\}$$

and there exists no $q' \in \underline{Q}$, $a' \in \underline{V} \cup \{B\}$,

$\delta \in \{L, R, 0\}$ such that

$(q, a) \to (q', a', \delta) \in \underline{I}\}$.

Less formally, S is obtained by adding to $\underline{I}$ a set of instructions transforming q into p for each pair (q, a) which does not occur as a left member of an instruction of $\underline{I}$.

It is clear that S halts if and only if it is in the state p and that $f_{S,n} = f_{T,n}$ for any $n \geq 1$.

3.8.3. Lemma 3.2: Let $\underline{V}$ be a vocabulary, ! and ? two symbols not in $\underline{V}$ and $\underline{W} = \underline{V} \cup \{!, ?\}$; for any pair (T, n) where T is a $\underline{V}$-TM and $n \geq 1$ is an integer, one may construct a $\underline{W}$-TM R such that $f_{R,n} = \{(x_1, x_2, \ldots, x_n, ! \, y \, ?) \mid (x_1, x_2, \ldots, x_n, y) \in f_{T,n}\}$ (*)

Proof.

The principle of the proof consists in constructing the $\underline{W}$-Turing machine R in such a way that its working is identical with that of the $\underline{V}$-Turing machine T except that at each moment the used fraction of the tape is embraced by the symbol pair ! , ? .

(*) Note that though R is a $\underline{W}$-Turing machine, the domain of $f_{R,n}$ is a subset of $\underline{V}^{*n}$: the symbols ! and ? only appear in the value of $f_{R,n}$.

If $T = (\underline{V}, \underline{Q}, \underline{I}, B, q_s)$, put $R = (\underline{W}, \underline{Q}_R, \underline{I}_R, B, q_{sR})$ with:

$\underline{Q}_R = (\underline{Q} \cup \{r, s, t\} \cup \{u_i \mid 1 \leq i \leq n+1\} \cup \{v_q \mid q \in \underline{Q}\} \cup \{w_q \mid q \in \underline{Q}\}$ (*)

$q_{sR} = r$

$\underline{I}_R$ is obtained by adding to $\underline{I}$ the following instructions:

$(r, a) \to (s, a, L)$ (1a)

for each $a \in \underline{V} \cup \{B\}$

$(s, B) \to (u_1, !, R)$ (1b)

$(u_i, a) \to (u_i, a, R)$ (2a)

for each $a \in \underline{V}$ and $1 \leq i \leq n$

$(u_i, B) \to (u_{i+1}, B, R)$ (2b)

for each $1 \leq i \leq n$

$(u_i, c) \to (u_i, c, 0)$ (3)

for each $c \in \{!, ?\}$ and $1 \leq i \leq n$

$(u_{n+1}, B) \to (t, ?, L)$ (4)

$(t, a) \to (t, a, L)$ (5a)

for each $a \in \underline{V} \cup \{B\}$

$(t, !) \to (q_s, !, R)$ (5b)

$(q, !) \to (v_q, B, L)$ (6a)

for each $q \in \underline{Q}$

$(v_q, B) \to (q, !, R)$ (6b)

for each $q \in \underline{Q}$

$(q, ?) \to (w_q, B, R)$ (7a)

for each $q \in \underline{Q}$

$(w_q, B) \to (q, ?, L)$ (7b)

for each $q \in \underline{Q}$

(*) It is of course supposed that the (notation used for the) states added to $\underline{Q}$ is different from that of the states of $\underline{Q}$.

The working of the $\underline{W}$-Turing machine R is illustrated by the following comments. The instructions (1a) through (5b) are applied just after the start of the Turing machine; their effect is to embrace the n arguments by the symbols ! and ?. More precisely, the instructions (1a) and (1b) write the symbol ! to the left of the first argument; the instructions (2a) and (2b) locate the symbol B following the nth argument; the instruction (3) warrants that the value of the function $f_{R,n}$ is undefined when one or more of the arguments is a string of $\underline{W}^* - \underline{V}^*$, i.e. warrants that the domain of $f_{R,n}$ is a subset of $\underline{V}^{*n}$; the instruction (4) writes the symbol ? to the right of the B following the nth argument; the instructions (5a) and (5b) shift the read-write device to its initial position and start the (simulated working of) the Turing machine T. The instructions (6a) and (6b) care for a correct working when the read-write device shifts off the used fraction of the tape to the left; the instructions (7a) and (7b) do the same for the case of a shift off to the right.

These comments clearly indicate that

$$f_{R,n} = \{(x_1, x_2, \ldots, x_n, ! \; y \; ?) \mid (x_1, x_2, \ldots, x_n, y) \in f_{T,n}\}.$$

3.8.4. Theorem 3.1: For any pair (T, n) where T is a Turing machine and $n \geq 1$ is an integer one may construct a Turing machine P which defines the same n-ary function as T and which is n-normal.

Proof.

If $T = (\underline{V}, \underline{Q}, \underline{I}, B, q_s)$ is the given $\underline{V}$-Turing machine, consider the $\underline{W}$-Turing machine $Q = (\underline{W}, \underline{Q}_Q, \underline{I}_Q, B, q_{sQ})$ obtained by applying to T successively the constructions of Lemma 3.2 and Lemma 3.1. From what precedes it should be clear that the Turing machine Q may be turned into an n-normal Turing machine defining the same n-ary function as the Turing machine T if one succeeds in "cleaning" the result with respect to the blank symbols and erasing the symbols ! and ?.

Let p be the state of $\underline{Q}_Q$, introduced by the construction of Lemma 3.1. Construct the $\underline{W}$-Turing machine $P = (\underline{W}, \underline{Q}_P, \underline{I}_P, B, q_{sQ})$ with:

$$\underline{Q}_P = \underline{Q}_Q \cup \{q_f, w, x\} \cup \{y_a \mid a \in \underline{V}\} \cup \{z_a \mid a \in \underline{V}\}$$

$\underline{I}_P$ is obtained by adding to $\underline{I}_Q$ the following instructions:

$$(p, a) \rightarrow (p, a, L) \quad \text{for each } a \in \underline{V} \cup \{B\} \tag{8a}$$

$$(p, !) \rightarrow (w, !, R) \tag{8b}$$

$$(w, B) \rightarrow (w, B, R) \tag{9a}$$

$$(w, a) \rightarrow (y_a, B, L) \quad \text{for each } a \in \underline{V} \tag{9b}$$

$$(w, ?) \rightarrow (x, B, L) \tag{10}$$

$$(y_a, B) \rightarrow (y_a, B, L) \quad \text{for each } a \in \underline{V} \tag{11a}$$

$$(y_a, c) \rightarrow (z_a, c, R) \quad \text{for each } a \in \underline{V} \text{ and } c \in \underline{V} \cup \{!\} \tag{11b}$$

$$(z_a, B) \rightarrow (w, a, R) \tag{12}$$

$$(x, a) \rightarrow (x, a, L) \quad \text{for each } a \in \underline{V} \cup \{B\} \tag{13a}$$

$$(x, !) \rightarrow (q_f, B, R) \tag{13b}$$

The working of the Turing machine P is illustrated by the following comments. The added instructions are applicable when the state p is entered, i.e. when the (simulated working of the) Turing machine Q halts. The instructions (8a) and (8b) locate the symbol !; the instructions (9a) and (9b) locate and erase the leftmost symbol on the right part of the tape; the instruction (10) erases the symbol ?; the instructions (11a) and (11b) locate the rightmost symbol on the left part of the tape while "carrying" the name of the symbol erased by the instruction (9b); the instruction (12) writes this symbol down; finally, the instructions (13a) and (13b) erase the symbol !.

From these comments it should be clear that the Turing machine P is n-normal and that $f_{P,n} = f_{T,n}$.

3.8.5. Corollary: An n-ary $\underline{V}$-string function is computable if and only if there exists an n-normal $\underline{W}$-Turing machine defining it, $\underline{W}$ being the vocabulary $\underline{V}$ extended with two supplementary symbols.

CHAPTER 4 : THE UNIVERSAL TURING MACHINE

Abstract: The notion of the universal V-Turing machine is introduced and the thesis of Turing is reformulated.

4.1. The string description of a Turing machine

4.1.1. Describing a Turing machine by a string over an extended vocabulary

Let $T = (\underline{V}, \underline{Q}, \underline{I}, B, q_s)$ be an arbitrary V-Turing machine. A little thought indicates that this Turing machine is univocally defined by the set $\underline{I}$ of its instructions, provided the blank symbol and the start state are represented by two known symbols.

Suppose now that the blank symbol is taken to be the particular symbol B, $B \notin \underline{V}$ (*). Provide the finite set $\underline{Q}$ with an arbitrary alphabetic order, say A_o, satisfying $A_o(q_s) = 1$. Replace each instruction of $\underline{I}$, say

$$(q, a) \rightarrow (q', a', \delta)$$

by the string

$$Q^{A_o(q)} a Q^{A_o(q')} a' \delta$$

where Q is an arbitrary symbol with $Q \notin \underline{V} \cup \{B\}$. Finally, concatenate in some arbitrary order the strings thus obtained and call e_T the resulting string. It should be clear that the string e_T univocally defines the Turing machine T - except for the exact names of the states (**); fortunately, this restriction is irrelevant as far as the working of a Turing machine is concerned, this working being independent from these names.

(*) cf. the footnote of Section 3.5.2.

(**) A knowledge of these names would require the knowledge of the alphabetic order A_o.

As a conclusion, if $\underline{V}$ is a vocabulary not containing the symbols B and Q, an arbitrary $\underline{V}$-Turing machine may be univocally described by a string, viz. e_T, over the (fixed) vocabulary $\underline{V} \cup \{B, Q, L, R, O\}$ (*). As an example, the string

QaQQaRQbQQbRQBQBOQQaQQBRQQbQQBRQQBQQQBO

univocally describes the {a, b}-Turing machine of Section 3.5.1.

In general, there are several strings e_T, describing a given Turing machine: they differ by the choice for the alphabetic order A_o and by the order in which the strings deduced from the instructions are concatenated.

On the other hand each string over the vocabulary $\underline{V} \cup \{B, Q, L, R, O\}$ is not a description e_T of a $\underline{V}$-Turing machine. To this end it has to be composed of substrings $Q^m a Q^n a' \delta$ with $m, n \geq 1$, $a, a' \in \underline{V} \cup \{B\}$, $\delta \in \{L, R, O\}$; moreover, the restriction (R) of Section 3.4.1 has to be satisfied. Clearly, it is possible to design a constructive procedure determining whether or not a given arbitrary string over $\underline{V} \cup \{B, Q, L, R, O\}$ is a description e_T of a $\underline{V}$-Turing machine.

4.1.2. Describing a Turing machine by a string over the original vocabulary

Let $T = (\underline{V}, \underline{Q}, \underline{I}, B, q_s)$, B, Q and e_T be defined as above. Provide now $\underline{V}$ with an arbitrary alphabetic order, say A. Moreover, put

$$\underline{V}_d = \underline{V} \cup \{B, Q, L, R, O\}$$

and provide this vocabulary with the alphabetic order B defined as follows:

$$\begin{aligned} B = A &\cup \{B, \text{card}(\underline{V}) + 1\} \cup \{Q, \text{card}(\underline{V}) + 2\} \\ &\cup \{L, \text{card}(\underline{V}) + 3\} \cup \{R, \text{card}(\underline{V}) + 4\} \\ &\cup \{O, \text{card}(\underline{V}) + 5\} \end{aligned} \qquad (**)$$

(*) The vocabularies $\underline{V} \cup \{B, Q\}$ and $\{L, R, O\}$ need not be disjoint.

(**) This implies that the vocabularies $\underline{V} \cup \{B, Q\}$ and $\{L, R, O\}$ are disjoint; if this is not the case, the symbols of the latter vocabulary should previously be "renamed" or, more precisely, replaced by symbols not in $\underline{V} \cup \{B, Q\}$.

Consider now the function

$$\underline{V}_d^*\text{-to-}\underline{V}^*$$

or, more precisely

$$(\underline{V}_d^*, B)\text{-to-}(\underline{V}^*, A).$$

The string over $\underline{V}$

$$d_T = \underline{V}_d^*\text{-to-}\underline{V}^*(e_T)$$

is called a *(V-)string description* of the Turing machine T.

As a conclusion an arbitrary $\underline{V}$-Turing machine is univocally described by a string over $\underline{V}$ (viz. the string description d_T) together with an alphabetic order in $\underline{V}$ (viz. A). In fact, from the alphabetic order A one may deduce the alphabetic order B; A and B univocally define the function $\underline{V}^*\text{-to-}\underline{V}_d^*$; further

$$e_T = \underline{V}^*\text{-to-}\underline{V}_d^*(d_T)$$

and, as indicated in Section 4.1.1, e_T univocally describes the $\underline{V}$-Turing machine T.

A Turing machine has in general different $\underline{V}$-string descriptions; they differ from each other by their correspondent strings e_T and/or by the alphabetic order in $\underline{V}$.

For the same reasons as above, each string over $\underline{V}$ is not a $\underline{V}$-string description of a $\underline{V}$-Turing machine. But again it is possible to design a constructive procedure determining whether or not an arbitrary given string over $\underline{V}$ is a $\underline{V}$-string description of a $\underline{V}$-Turing machine; in other words, if a is an arbitrary given symbol of $\underline{V}$, the $\underline{V}$-string function S defined by:

$$S : \underline{V}^* \to \{a, \varepsilon\} :$$

$$S(x) = \begin{cases} a & \text{if there exists a } \underline{V}\text{-Turing machine such that x is a } \underline{V}\text{-string description of it;} \\ \varepsilon & \text{otherwise} \end{cases}$$

is computable (*).

(*) The choice of a and ε is arbitrary: any two other strings of $\underline{V}^*$ would do as well. Note that S is a predicate if a and ε are interpreted as true and false respectively.

4.2. The universal Turing machine

The procedure (i) to (v) of Section 3.4.6 simulates the working of a given $\underline{V}$-Turing machine T calculating the value of an n-ary function. Actually, this procedure may be made to simulate the working of any $\underline{V}$-Turing machine T calculating the value of an n-ary function: to this effect (a description of) the Turing machine T should be presented to the procedure in the form of a (supplementary) argument. More precisely, consider the following procedure to which the (n+1)-tuple $(w, x_1, x_2, \ldots, x_n)$ of $(\underline{V}^*)^{n+1}$ is presented:

(i) compute S(w) (*);
if S(w) = ε, go to step (i) (**); if S(w) = a, proceed;

(ii) determine the $\underline{V}$-Turing machine T of which w is a $\underline{V}$-string description;

(iii) perform steps (i) to (v) of the procedure of Section 3.4.6, being understood that the Turing machine to be considered is the one determined during step (ii) above.

It should be clear that this procedure is constructive and that it simulates the working of an arbitrary $\underline{V}$-Turing machine, a $\underline{V}$-string description of which is presented to it in the form of an argument. According to the thesis of Turing there exists a $\underline{V}$-Turing machine, say U, which defines the same $\underline{V}$-string function as this procedure, i.e. such that:

$$f_{U,n+1} = \{(w, x_1, x_2, \ldots, x_n, y) \mid S(w) = a \text{ and } (x_1, x_2, \ldots, x_n, y) \in f_{T,n} \text{ where T is the } \underline{V}\text{-Turing machine of which w is a } \underline{V}\text{-string description } (^{***})\}. \quad (1)$$

As a conclusion, for a given vocabulary $\underline{V}$ (provided with an alphabetic order A) there exists a $\underline{V}$-Turing machine U which "simulates" the working of any other $\underline{V}$-Turing machine calculating the value of an n-ary function, in the sense that $f_{U,n+1}$ satisfies the equality (1). This Turing machine U is called the *universal ($\underline{V}$-)Turing machine* (****).

(*) S is the function defined in Section 4.1.2; $\underline{V}$ is supposed to be provided with a known alphabetic order A.

(**) i.e. enter an infinite loop.

(***) The existence of this Turing machine is implied by S(w) = a.

(****) It is more correct to speak of "a" (rather than "the") universal Turing machine: if there exists a Turing machine defining a function, there exists an infinity of them.

4.3. Discussion

The universal Turing machine is similar to a (general purpose) computer: its first argument corresponds to the program (calculating the function sine, for instance), its other arguments to the data (30°, for instance) for which the calculation defined by the program has to be performed.

The definition of a computable function of Section 3.6 may now be reformulated as follows:

an n-ary $\underline{V}$-string function F is computable if there exists a string w, $w \in \underline{V}^*$ such that

$$F = \{(x_1, x_2, \ldots, x_n, y) \mid (w, x_1, x_2, \ldots, x_n, y) \in f_{U,n+1}\}$$

The thesis of Turing may be restated in a similar way; as a result this thesis hinges on one machine (viz. the universal Turing machine) only.

CHAPTER 5 : SOME FUNCTIONS WHICH ARE NOT COMPUTABLE

Abstract: It is shown that there exist functions which are not computable. Some of them are of practical interest in computer programming.

5.1. The halting problem

5.1.1. Introduction

Suppose one wants to calculate the value of the 1-ary (*) function $f_{T,1}$ defined by a V-Turing machine T for a given argument x.

Before starting the Turing machine it would be interesting to know whether it will ever halt or not, i.e. whether $f_{T,1}(x)$ is defined or not; in fact, in the latter case it is not worthwhile to start it. It would thus be interesting to be able to calculate the value of the 2-ary function H defined as follows: its arguments are respectively a (string description of the) Turing machine T, and x; its value is some symbol of V, say a, if $f_{T,1}(x)$ is defined, and ε if $f_{T,1}(x)$ is undefined (**).

It is the goal of this section to prove that the V-string function H is not computable.

5.1.2. The function H

For a given vocabulary V the function H may be defined as follows: provide V with an arbitrary alphabetic order, let a be an arbitrary symbol of V and S be defined as in Section 4.1.2; then H is the 2-ary V-string function

$$H : \{ x \mid S(x) = a\} \times V^* \rightarrow \{a, \varepsilon\}:$$

$$H(x, y) = \begin{cases} a \text{ if } f_{T,1}(y) \text{ is defined, T being the V-Turing machine of which x is a string description;} \\ \varepsilon \text{ if } f_{T,1}(y) \text{ is undefined, T being the V-Turing machine of which x is a string description.} \end{cases}$$

(*) For the sake of simplicity only 1-ary functions are considered.

(**) cf. the last footnote of Section 4.1.2.

Less formally:

$$H\ (d_T,\ y) = \begin{cases} a \text{ if } f_{T,1}(y) \text{ is defined} \\ \varepsilon \text{ if } f_{T,1}(y) \text{ is undefined.} \end{cases}$$

5.1.3. The proof

The principle of the proof consists in assuming the function H is computable and in showing that this hypothesis leads to a contradiction.

If H is computable there exists a $\underline{V}$-Turing machine, say h, which defines H, i.e. such that

$$f_{h,2} = H$$

Introduce now the 1-ary $\underline{V}$-string function J:

$$J : \{\ x \mid S(x) = a\ \} \rightarrow \{a,\ \varepsilon\} : J(x) = H(x,\ x).$$

This function is identical with H except that it restricts its attention to the case in which both arguments of H are equal. The function J provides an answer to the question whether or not an arbitrarily given $\underline{V}$-Turing machine T halts when it is started with the argument d_T; in other words J restricts its attention to the working of a Turing machine which is started with its own description on the tape! It is clear that the value of J is of theoretical interest only. Clearly, J is computable because H is: a Turing machine j defining J may be identical with h except that it first makes a copy of its argument (i.e. that it first replaces the contents xB of its tape by xBxB).

Introduce now the 1-ary $\underline{V}$-string function K:

$$K : \{\ x \mid J(x) = \varepsilon\ \} \rightarrow \{\varepsilon\} : K(x) = J(x).$$

This function is identical with J except that it is undefined when the value of J is a. Again, K is computable because J is: the $\underline{V}$-Turing machine k defining K is identical with j except that, when j halts, k determines whether the result is a or ε; in the former case it enters an infinite loop, in the latter case it halts.

Let d_k be a $\underline{V}$-string description of the $\underline{V}$-Turing machine k. Question: is $K(d_k)$ defined or not?

Suppose first that $K(d_k)$ is not defined, i.e. that the $\underline{V}$-Turing machine k does not halt after having been started with the string d_k on its tape. Then, by definition of the function H:

$$H(d_k,\ d_k) = \varepsilon$$

Hence

$$J(d_k) = \varepsilon$$

and

$$K(d_k) = \varepsilon$$

i.e. $K(d_k)$ is defined; this contradicts the assumption that $K(d_k)$ is not defined.

Suppose now that $K(d_k)$ is defined, i.e. that the $\underline{V}$-Turing machine k halts after having been started with d_k. Now

$$H(d_k, d_k) = a$$

$$J(d_k) = a$$

and $K(d_k)$ is undefined. Again this contradicts the assumption that $K(d_k)$ is defined.

Hence both assumptions lead to a contradiction; this is impossible because $K(d_k)$ is either defined or not. Hence the hypothesis that H is a computable function can not be verified. This proves the theorem.

5.1.4. Remark

It was just proved that there does not exist a Turing machine h which for each arbitrarily given $\underline{V}$-Turing machine T and each arbitrarily given string x in $\underline{V}^*$ determines whether or not T halts after having been started with x.

But this does not exclude the existence of some Turing machine h' which for each arbitrarily given $\underline{V}$-Turing machine T out of some limited class of $\underline{V}$-Turing machines and/or each arbitrarily given string x of some subset of $\underline{V}^*$ determines whether or not T halts after having been started with x. For instance, if our attention is limited to those $\underline{V}$-Turing machines which contain only one instruction, it is possible to construct the corresponding Turing machine h'; equivalently, it is possible to design a constructive procedure which for an arbitrarily given $\underline{V}$-Turing machine of this class and an arbitrarily given string x of $\underline{V}^*$ determines whether or not T halts after having been started with x; in fact, if $(q, a) \rightarrow (q', a', \delta)$ is the (single) instruction of a $\underline{V}$-Turing machine, it does not halt after having been started with x if and only if

$$q = q', \; a = a', \; \delta = 0, \; x \in \{a\}.\underline{V}^*$$

or: $$q = q', \; a = a' = B, \; \delta \in \{R,L\}, \; x = \varepsilon.$$

Another example is obtained if our attention is limited to the class consisting of the (single) Turing machine of Section 3.5.1; in fact, this Turing machine halts if and only if the string presented to it is not ε.

As a conclusion the non-existence of the Turing machine h is related to the generality of the function it has to define or, equivalently, to the extent of the domain of this function; by restricting the domain in such a way as to exclude all "nasty" cases (such as the case of the function K with the argument d_k) one yields a computable function.

5.1.5. Another terminology

The problem which consists in calculating the value of a total function is said to be *(un)solvable* (or: *(un)decidable*) when the function is (not) computable. For instance, one says that the halting problem for a Turing machine is unsolvable.

5.1.6. A practical aspect

The fact that it is not possible to construct the Turing machine h may - thanks to Turing's thesis - also be formulated as follows: it is not possible to write a computer program which for an arbitrarily given other program, say P, determines whether or not the execution time of P is finite. Note that this is not equivalent to saying that it is not possible to write a computer program determining whether or not an arbitrarily given other program enters an infinite loop during its execution: when a program does not halt, this is not necessarily because it enters an infinite loop (*).

5.2. The blank tape halting problem

5.2.1. Introduction

The halting problem treated above appeared to be too general to be solvable (see also Section 5.1.4). Therefore one may hope to be able to solve the following, less general, problem: determine whether or not an arbitrarily given V-Turing machine halts after having been started with a blank (**) tape.

(*) While the notion "entering an infinite loop" has never been exactly defined, it suggests a behaviour of some cyclic nature (see e.g. Section 3.5.1).

(**) i.e. containing a string of B's.

More precisely, the function to be studied now is a 1-ary function H'; the argument of this function is a string description of a V-Turing machine T; the value of the function is a when $f_{T,1}(\varepsilon)$ is defined, ε when $f_{T,1}(\varepsilon)$ is undefined. (*)

This problem is called the blank tape halting problem and is less general than the halting problem in that only the V-Turing machine is arbitrary, the initial contents of the tape being fixed. Nevertheless, the function H' is not a computable function neither, as will now be shown.

5.2.2. Proof

The principle of the proof consists in showing that if the blank tape halting problem were solvable the halting problem would be solvable too.

Associate with each pair (T, x) where T is a V-Turing machine and x is a string of V^*, a V-Turing machine T_x. This Turing machine T_x is identical with T except that after having been started it first writes the string x on its tape - thus (partially) overwriting its initial contents. Hence, except for some preliminary operations, the working of T_x started with a blank tape is identical with the working of T started with x. In particular, T_x halts when started with a blank tape if and only if T halts when started with the string x on its tape.

Assume now that the blank tape halting problem is solvable. The halting problem is then solvable too. In fact, suppose it has to be determined whether or not an arbitrarily given V-Turing machine T halts when started with an arbitrarily given string x on its tape. The following constructive procedure would yield the answer: construct the Turing machine T_x associated with (T, x) as indicated above; determine whether or not T_x halts when started with a blank tape (this is possible because the blank tape halting problem is supposed to be solvable); the answer found is the answer to the question whether or not T halts when started with x.

5.3. The uniform halting problem

5.3.1. Introduction

The problem is now: determine whether or not an arbitrarily given V-Turing machine halts whatever be the string of V^* on its tape at the start or, in other words, determine for an arbitrarily given V-Turing machine T whether or not the V-string function $f_{T,1}$ is total.

(*) Note that $f_{T,1}(\varepsilon)$ is defined if and only if $f_{T,2}(\varepsilon, \varepsilon)$, $f_{T,3}(\varepsilon, \varepsilon, \varepsilon)$, ... are defined.

The function to be studied now is a 1-ary function H"; its argument is a string description of a V-Turing machine T; its value is a when $f_{T,1}$ is total, ε otherwise. Again, H" will be shown not to be computable.

5.3.2. Proof

The proof consists in showing that if the uniform halting problem were solvable the blank tape halting problem would be solvable too.

To this end associate with each V-Turing machine T a V-Turing machine S identical with T except that after having been started it first erases the contents of the tape; hence the (second part of the) working of S started with an arbitrary tape contents is identical with the working of T started with a blank tape. By the way, S is a rather "uninteresting" Turing machine: the domain of the function $f_{S,1}$ is either the empty set or $\underline{V}^*$; moreover, in the latter case its value is constant: $f_{S,1}(x) = f_{T,1}(\varepsilon)$ for any $x \in \underline{V}^*$. Anyhow, the function $f_{S,1}$ is total if and only if the Turing machine T halts when started with a blank tape.

Assume now that the uniform halting problem is solvable. The blank tape halting problem is then solvable too. In fact, suppose it has to be determined whether or not an arbitrarily given V-Turing machine T halts when started with a blank tape. The following constructive procedure would yield the answer: construct the Turing machine S associated with T; determine whether or not $f_{S,1}$ is total; the answer is also the answer to the question whether or not T halts when started with a blank tape.

5.4. The equivalence problem

5.4.1. Introduction

Suppose F is a non-empty (*) computable 1-ary V-string function. The problem is: determine for an arbitrarily given V-Turing machine T whether or not $f_{T,1} = F$.

In other words, the function to be studied is a 1-ary function H_F; its argument is a string description of a V-Turing machine T; its value is a when $f_{T,1} = F$, ε otherwise. Again, H_F is not computable.

(*) This excludes the trivial case of the function F = {}, i.e. the function the value of which is undefined for any argument.

5.4.2. Proof

The proof consists in showing that the solvability of the equivalence problem for the function F implies the solvability of the blank tape halting problem.

As F is a computable V-string function there exists a V-Turing machine R such that $f_{R,1} = F$. Associate now with each V-Turing machine T a V-Turing machine P the working of which, when started with x, consists of two parts executed successively: the first part is identical with the working of T when started with a blank tape, the second part is identical with the working of R when started with x (*). Clearly, $f_{P,1} = F$ if $f_{T,1}(\varepsilon)$ is defined, and $f_{P,1} = \{\} \; (\neq F)$ otherwise.

Assume now that the equivalence problem for the function F is solvable. The blank tape halting problem is then solvable too. In fact, suppose it has to be determined whether or not an arbitrarily given V-Turing machine T halts when started with a blank tape. Construct the associated Turing machine P and determine whether or not $f_{P,1} = F$. The answer is also the answer to the question whether or not T halts when started with a blank tape.

5.5. General remark

The fact that there exist unsolvable problems may be looked at from two different points of view.

From a "pragmatic" point of view a problem is unsolvable because it is too general to allow a solution. Actually, a problem is unsolvable when it includes peculiar cases such as the one considered in the proof of Section 5.1.3. Hence, if one restricts the problem such as to include only "normal" cases it should become solvable. These "normal" cases certainly encompass, for instance, most of the "daily-life" computer programs because a programmer is generally able to prove, for instance, whether or not his program halts. Though this argument suggests that

---

(*) Note that during the first part the tape contents x have to be "saved" in order to be available at the start of the second part. Though this may put some "organizational" problems during the design of the Turing machine P, it may be constructed: in fact, it is clear that there exists a constructive procedure simulating the working of P.

the existence of unsolvable problems is of pure theoretical interest, it appears to be very difficult to clearly define what a "normal" case is; more precisely, the numerous fruitless results of the last decade seem to indicate that it is very difficult to restrict an unsolvable problem - such as the halting problem - in such a way that it becomes solvable while still covering a reasonable amount of non-trivial cases.

From a "philosophical" point of view a problem is unsolvable (i.e. the corresponding Turing machine does not exist) because the theory is too "narrow". This is comparable with the situation occurring in the theory of natural numbers in which, for instance, the square root of 2 "does not exist". The solution then lies in a "generalization" of the theory, e.g. in the replacement of the theory of natural numbers by the theory of real numbers. Unfortunately, in a thus generalized theory of computable functions the formal notion of computability would no longer correspond with our intuitive notion of computability. Post expressed this fact by saying that the existence of non-computable functions indicates the existence of a "limitation of the mathematicizing power of the homo sapiens".

CHAPTER 6 : EFFECTIVELY ENUMERABLE AND DECIDABLE SETS

Abstract: Effectively enumerable and decidable sets are defined and some of their properties indicated.

6.1. Introduction

This chapter is concerned with sets of strings over a given vocabulary $\underline{V}$. With the help of the notion of computable function two classes of sets are defined: the effectively enumerable sets and the decidable sets.

The properties of effectively enumerable and of decidable sets shed another light on the notion of computability. They are of direct use in the theory of formal languages and, as a consequence, in mathematical logic and in the theory of programming languages.

6.2. Definitions

6.2.1. Acceptor and decision function of a set

Let $\underline{V}$ be a vocabulary and a one of its symbols.

With any set $\underline{S}$ of strings over $\underline{V}$ are associated two $\underline{V}$-string functions defined as follows.

The *acceptor function* of the set $\underline{S}$ is the 1-ary $\underline{V}$-string function

$$A_S : \underline{S} \rightarrow \{a\} : A_S(x) = a$$

The particularity of this function is that its value is undefined for any argument belonging to the complement of $\underline{S}$.

The *decision function* of the set $\underline{S}$ is the 1-ary $\underline{V}$-string function

$$D_S : \underline{V}^* \rightarrow \{a, \varepsilon\} : D_S(x) = \begin{cases} a \text{ if } x \in \underline{S} \\ \varepsilon \text{ if } x \in \underline{V}^* - \underline{S} \end{cases}$$

As a difference with the acceptor function, the decision function is a total $\underline{V}$-string function.

6.2.2. Effective enumerability and decidability

A set of strings over $\underline{V}$ is *effectively enumerable* when its acceptor function is computable (*) (**).

A set of strings over $\underline{V}$ is *decidable* when its decision function is computable (*).

Most sets one may think of are effectively enumerable as well as decidable. For instance, it is easy (and even trivial!) to design a constructive procedure which defines the acceptor or the decision function of the set $\{a^n b^n \mid n \geq 1\}$ over the vocabulary $\{a, b\}$.

6.3. Effectively enumerable sets and the domain of computable functions

Theorem 6.1: A set (of strings over a vocabulary $\underline{V}$) is effectively enumerable if and only if it is the domain of a partial computable (1-ary $\underline{V}$-string) function.

Proof.

Suppose that $\underline{S}$ is the domain of the computable 1-ary $\underline{V}$-string function F. As F is computable there exists a $\underline{V}$-Turing machine T such that $f_{T,1} = F$. Consider now the Turing machine R identical with T except for the following: when T halts, R replaces the contents of the tape by the symbol a. Clearly, $f_{R,1} = A_S$, i.e. the Turing machine R defines the acceptor function of the set $\underline{S}$. This proves the first part of the theorem.

Suppose now that $\underline{S}$ is an effectively enumerable set. Then it is the domain of its acceptor function which is a computable 1-ary $\underline{V}$-string function. This proves the second part of the theorem.

6.4. Effectively enumerable sets and the range of total computable functions

6.4.1. Lemma 6.1: Let $\underline{S}$ be an effectively enumerable set of strings over a vocabulary $\underline{V}$ and d a symbol, $d \notin \underline{V}$; one may associate with this set a total computable 1-ary $(\underline{V} \cup \{d\})$-string function $F : \underline{V}^* \twoheadrightarrow \underline{S} \cup \{d\}$.

(*) In older literature the terms "recursively enumerable" and "recursive" replace "effectively enumerable" and "decidable".

(**) The term "effectively enumerable" should not be confused with the term "denumerable" occurring in set theory.

Proof.

Provide $\underline{V}$ with an alphabetic order, say A.

Consider the 1-ary ($\underline{V} \cup \{d\}$)-string function F defined by the following constructive procedure which makes correspond with each string of $\underline{V}^*$, say x, a string over $\underline{V} \cup \{d\}$:

(i) determine the strings $y, z \in \underline{V}^*$ such that

$$x = \underline{V}^{*2}\text{-to-}\underline{V}^*(y, z) \qquad (^*) \qquad ;$$

more precisely, compute

$$y = \overline{\underline{V}^{*2}\text{-to-}\underline{V}^*}^{(1)}(x) \qquad (^{**})$$

$$z = \overline{\underline{V}^{*2}\text{-to-}\underline{V}^*}^{(2)}(x) \qquad (^{**})$$

(ii) compute the non-negative integer m defined by:

$$m = \underline{V}^*\text{-to-}\underline{N}(z) \qquad (^{**})$$

(iii) let T be a $\underline{V}$-Turing machine defining the acceptor function of the (effectively enumerable) set $\underline{S}$ and q_s and B its start state and blank symbol respectively; start T with the initial instantaneous description

$$(q_s, \varepsilon, yB)$$

and (try to) apply successively m instructions;

(iv) distinguish the three following cases:

(1°) it is not possible to apply m instructions because a final instantaneous description is reached before; put F(x) = d;

(2°) the instantaneous description obtained after having applied m instructions is a final one; put F(x) = y;

(3°) the instantaneous description obtained after having applied m instructions is not a final instantaneous description; put F(x) = d;

(v) stop.

(*) By $\underline{V}^{*2}\text{-to-}\underline{V}^*$ is meant the function $(\underline{V}^{*2}, A)\text{-to-}(\underline{V}^*, A)$.

(**) Remember that these functions are computable (section 3.7).

It is clear that the domain of the 1-ary function F defined by this procedure is $\underline{V}^*$; in fact, the procedure comes to an end for any string x, $x \in \underline{V}^*$. The proof is now completed if the range of F is shown to be $\underline{S} \cup \{d\}$.

If $y_o \neq d$ is a string of the range of F, it is so by virtue of the case (2°) of step (iv). Hence it is a string of $\underline{S}$ by virtue of the definition of the acceptor function.

Conversely, if y_o is a string of $\underline{S}$, the Turing machine T will halt after the application of a finite number, say m_o, of instructions after having been started with (q_s, ε, y_oB). Hence the procedure described will deliver y_o as the value of the function F for the argument

$$\underline{V}^{*2}\text{-to-}\underline{V}^*(y_o, \underline{N}\text{-to-}\underline{V}^*(m_o))$$

6.4.2. Interpretation of the preceding lemma

Lemma 6.1 asserts that the elements of an effectively enumerable set may all be obtained (i.e. may be "enumerated") by calculating successively

$$F(\underline{N}\text{-to-}\underline{V}^*(0)),\ F(\underline{N}\text{-to-}\underline{V}^*(1)),\ F(\underline{N}\text{-to-}\underline{V}^*(2)),\ \ldots$$

This enumeration is "effective" in the sense that the calculation of any value F(x) of F takes only a finite amount of time. This property is at the origin of the name "effective enumerability". (*)

Note that from a practical point of view the enumeration of the elements of $\underline{S}$ with the help of the function F has a bad efficiency: most of the values of F are the symbol d and the acquisition of the elements of $\underline{S}$ proceeds slowly. This is related to the fact that the arguments x of F are in some sense in a one-to-one correspondence with the instantaneous descriptions of T (through y and m), and that only such arguments deliver a string of $\underline{S}$ which correspond with a final instantaneous description.

(*) By the way,

$$\underline{N}\text{-to-}\underline{V}^*(0),\ \underline{N}\text{-to-}\underline{V}^*(1),\ \underline{N}\text{-to-}\underline{V}^*(2),\ \ldots$$

is an effective enumeration of the elements of $\underline{V}^*$.

6.4.3. Lemma 6.2: If a set $\underline{S}$ of strings over $\underline{V}$ is the range of a total computable 1-ary $\underline{V}$-string function $F : \underline{V}^* \twoheadrightarrow \underline{S}$ then $\underline{S}$ is effectively enumerable.

Proof.

Suppose x is an arbitrarily given string of $\underline{V}^*$.

Consider now the following procedure: calculate successively

$$F(\underline{N}\text{-to-}\underline{V}^*(0)),\ F(\underline{N}\text{-to-}\underline{V}^*(1)),\ F(\underline{N}\text{-to-}\underline{V}^*(2)),\ \ldots$$

until a value equal to x is obtained; in that case, stop with the symbol a as result.

It is clear that this procedure defines the acceptor function of the set $\underline{S}$ (*).

6.4.4. Theorem 6.2: A non-empty set of strings over $\underline{V}$ is effectively enumerable if and only if it is the range of a non-empty total computable 1-ary $\underline{V}$-string function.

Proof.

The "if" part of the theorem directly follows from Lemma 6.2.

In order to prove the "only if" part, let $\underline{S}$ be a non-empty effectively enumerable set and F the function of Lemma 6.1. Consider the 1-ary function E defined by the following procedure which makes correspond with each string x of $\underline{V}^*$ another string of $\underline{V}^*$:

(i) calculate successively

$$F(\underline{N}\text{-to-}\underline{V}^*(0)),\ F(\underline{N}\text{-to-}\underline{V}^*(1)),\ F(\underline{N}\text{-to-}\underline{V}^*(2)),\ \ldots$$

until a value different from d is obtained; call x_o this value;

(ii) put $E(x) = \begin{cases} F(x) & \text{if } F(x) \neq d \\ x_o & \text{if } F(x) = d \end{cases}$

(iii) stop.

It is easy to see that E is a function $E : \underline{V}^* \twoheadrightarrow \underline{S}$; in particular, the execution of step (i) takes only a finite amount of time as by hypothesis $\underline{S}$ is a not-empty set, i.e. the string x_o exists.

(*) Note in particular that F is total and that, as a consequence, the computation time of each value of F is finite.

6.5. A set which is not effectively enumerable

6.5.1. The set

The goal of this section is to prove that there exist sets which are not effectively enumerable.

To this end, the following set (of strings over a vocabulary $\underline{V}$ provided with an alphabetic order and containing the symbol a) will be proved not to be effectively enumerable:

$$\underline{P} = \{ x \mid S(x) = a \text{ and } f_{T,1} \text{ is total, T being the } \underline{V}\text{-Turing machine of which x is a } \underline{V}\text{-string description} \} \quad (^*)$$

Less formally, $\underline{P}$ is the set of the Turing machines defining a total 1-ary function.

6.5.2. The proof

Assume that $\underline{P}$ is effectively enumerable. According to Theorem 6.2 there exists a total computable 1-ary $\underline{V}$-string function E the range of which is $\underline{P}$ (**).

Consider now the 1-ary $\underline{V}$-string function D defined by

$$D : \underline{V}^* \to \underline{V}^* : D(x) = \text{conc }(a, f_{U,2}(E(x), x)) \qquad (1)$$

where conc is the concatenation function (see Section 2.2) and U the universal $\underline{V}$-Turing machine (***); that D is a total function directly follows from the fact that E is a total function and that E(x) is a string description of a $\underline{V}$-Turing machine defining a total 1-ary function.

Now, as D is a computable (****) $\underline{V}$-string function there exists a $\underline{V}$-Turing machine R which defines D. Let d_R be a $\underline{V}$-string description of it; by definition of the universal $\underline{V}$-Turing machine U any string x of the domain of D, i.e. any $x \in \underline{V}^*$, satisfies:

$$D(x) = f_{U,2}(d_R, x) \qquad (2)$$

(*) S is the function of Section 4.1.2.

(**) It is evident that $\underline{P}$ is not empty.

(***) The notation conc (a, $f_{U,2}(...)$) is used instead of $af_{U,2}(...)$ for more clarity.

(****) That D is computable is evident: E, $f_{U,2}$ and conc are.

But as R is a V-Turing machine defining a total 1-ary function its V-string descriptions and, in particular, the string d_R must be elements of P. Hence, by definition of the function E there must be a string $x_o \in V^*$ such that

$$E(x_o) = d_R$$

and (2) may be written

$$D(x) = f_{U,2}(E(x_o), x) \qquad \text{for all } x \in V^* \tag{3}$$

From (1) and (3) it follows that for all $x \in V^*$ the following equality holds:

$$f_{U,2}(E(x_o), x) = \text{conc}(a, f_{U,2}(E(x), x))$$

This leads to a contradiction for $x = x_o$ because for any string y it cannot be the case that y = conc (a, y). Hence the hypothesis that P is effectively enumerable cannot be true.

6.5.3. Corollary: The set of all effectively enumerable sets of strings over a vocabulary V is a proper subset of the set of all sets of strings over V.

6.5.4. Comment

The principle of the proof of Section 6.5.2 may be illustrated by the following (very informal!) proof of a similar property for non-negative integers.

Suppose it is possible to enumerate all total computable 1-ary N-functions, for instance as

$$f_o, f_1, f_2, \ldots$$

Consider now the function

$$d(n) = 1 + f_n(n) \tag{a}$$

As d(n) is total it must appear somewhere in the list, i.e. for some n_o

$$d(n) = f_{n_o}(n) \tag{b}$$

The equalities (a) and (b) lead to

$$f_{n_o}(n) = 1 + f_n(n)$$

and, for $n = n_o$:

$$f_{n_o}(n_o) = 1 + f_{n_o}(n_o)$$

6.6. Decidable sets versus effectively enumerable sets

6.6.1. Lemma 6.3: A decidable set (of strings over a given vocabulary $\underline{V}$) is effectively enumerable.

Proof.

A Turing machine T_D defining the decision function of the set may be turned into a Turing machine T_A defining the acceptor function of the set by the following modification: when T_D halts with the result ε, T_A enters an infinite loop.

6.6.2. Lemma 6.4: The complement (with respect to $\underline{V}^*$) of a decidable set (of strings over a given vocabulary $\underline{V}$) is effectively enumerable.

Proof.

A Turing machine T_D defining the decision function of the set may be turned into a Turing machine T'_A defining the acceptor function of the complement of the set by the following modifications: when T_D halts with the result a, T'_A enters an infinite loop; when T_D halts with the result ε, T'_A halts with the result a.

6.6.3. Lemma 6.5: If a set $\underline{S}$ (of strings over $\underline{V}$) and its complement (with respect to $\underline{V}^*$) are both effectively enumerable, the set $\underline{S}$ is decidable.

Proof.

Let T_A and T'_A be two $\underline{V}$-Turing machines defining the acceptor function of $\underline{S}$ and $\underline{V}^* - \underline{S}$ respectively.

Consider now the following procedure to which an arbitrary string x of $\underline{V}^*$ is given:

(i) prepare T_A to start the acceptance test of the string x, i.e. make T_A have the instantaneous description (q_s, ε, xB) (*);

(ii) idem as step (i) for T'_A;

(*) q_s and B are the start state and the blank symbol of T_A.

(iii) apply an instruction to T_A; if the resulting instantaneous description is a final one, stop with the result a;

(iv) apply an instruction to T'_A; if the resulting instantaneous description is a final one, stop with the result ε;

(v) go to step (iii).

It is clear that this procedure defines the decision function of the set $\underline{S}$; note in particular that it halts after a finite time: x is an element either of $\underline{S}$ or of $\underline{V}^* - \underline{S}$ and hence either T_A or T'_A lead to a final instantaneous description after a finite time.

6.6.4. Theorem 6.3: A set $\underline{S}$ (of strings over a vocabulary $\underline{V}$) is decidable if and only if $\underline{S}$ and its complement (with respect to $\underline{V}^*$) is effectively enumerable.

Proof.

From Lemma 6.3, 6.4 and 6.5.

6.7. An effectively enumerable set which is not decidable

6.7.1. The set

The goal of this section is to show on the hand of an example that there exist effectively enumerable sets which are not decidable.

The following set of strings (over a vocabulary $\underline{V}$ provided with an alphabetic order and containing the symbol a) will be proved to be effectively enumerable while not being decidable:

$\underline{Q}$ = { x | there exist y, z $\in \underline{V}^*$ such that x = $\underline{V}^{*2}$-to-$\underline{V}^*$(y, z), S(y) = a and $f_{T,1}(z)$ is defined, T being the $\underline{V}$-Turing machine of which y is a $\underline{V}$-string description}.

Less formally, $\underline{Q}$ is equivalent with the set of pairs (d_T, z) such that $f_{T,1}(z)$ is defined.

6.7.2. The proof

Consider the following procedure to which an arbitrary string x of $\underline{V}^*$ is presented:

(i) calculate the strings y and z defined by:

$$y = \overline{\underline{V}^{*2}\text{-to-}\underline{V}^{*}}^{(1)}(x)$$

$$z = \overline{\underline{V}^{*2}\text{-to-}\underline{V}^{*}}^{(2)}(x)$$

(ii) calculate the value S(y); if S(y) = ε, go to step (ii) (*);

(iii) start the V-Turing machine T of which y is a V-string description, with the string z on its tape; if, after some time, this Turing machine halts, stop with the result a.

It is clear that this procedure defines the acceptor function of Q; note in particular that step (iii) does not terminate if and only if $f_{T,1}(z)$ is undefined. Hence the set Q is effectively enumerable.

Consider now the following procedure to which an arbitrary pair $(y, z) \in V^{*2}$ is presented:

(i) calculate S(y); if S(y) = ε, go to step (i);

(ii) calculate the value

$$D_Q(V^{*2}\text{-to-}V^*(y, z))$$

where D_Q is the decision function of the set Q;

(iii) stop with the value obtained during step (ii) as a result.

Clearly, this procedure defines the function H of Section 5.1.2. If the set Q were decidable, this procedure would be constructive and, consequently, the function H would be computable; less formally, the decidability of Q would imply the solvability of the halting problem. Hence the set Q is not decidable.

6.7.3. Corollary: The set of all decidable sets of strings over a vocabulary V is a proper subset of the set of all effectively enumerable sets of strings over V.

6.7.4. Note

The set $V^* - Q$ where Q is defined as in Section 6.7.2 is not effectively enumerable. This directly follows from Theorem 6.3.

(*) i.e. enter an infinite loop.

6.8. Some informal comments

6.8.1. A practical aspect

When a set is decidable it is possible to determine whether or not an arbitrarily given string is one of its elements. This determination may be done for instance by a Turing machine, or by hand according to a procedure, or by a computer through execution of a program.

When a set is effectively enumerable its sentences may be enumerated (by a Turing machine, by hand, by a computer) but is is not necessarily possible to determine whether or not an arbitrarily given string is an element of this set. More precisely, when a set is effectively enumerable without being decidable it is possible to decide (with the help of a Turing machine, by hand, by a computer) that an arbitrarily given string is an element of the set, if such is the case; but it is in general not possible to decide that an arbitrarily given string is not an element of this set: in fact, as long as the Turing machine defining the acceptor function did not halt one does not know whether it will halt at some later time or never halt.

A programming language may be considered as the set of all correct programs written in this language. Clearly, it is highly desirable that such a language be decidable in order to be able to detect the incorrect programs, i.e. the programs containing one or more (syntactical) errors.

6.8.2. A graphical representation

The results of Section 6.5 and Section 6.7 may be represented graphically as follows:

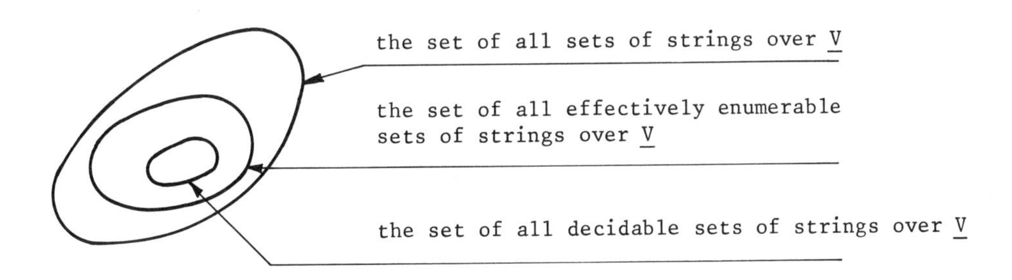

6.8.3. On the nature of effective enumerability and decidability

A subset of a decidable set is not necessarily decidable: such a subset may even be not effectively enumerable! For instance, any non effectively enumerable set of strings over $\underline{V}$, such as the set $\underline{P}$ of Section 6.5.1, is a proper subset of the decidable set $\underline{V}^*$ (*). This situation may be represented graphically as follows:

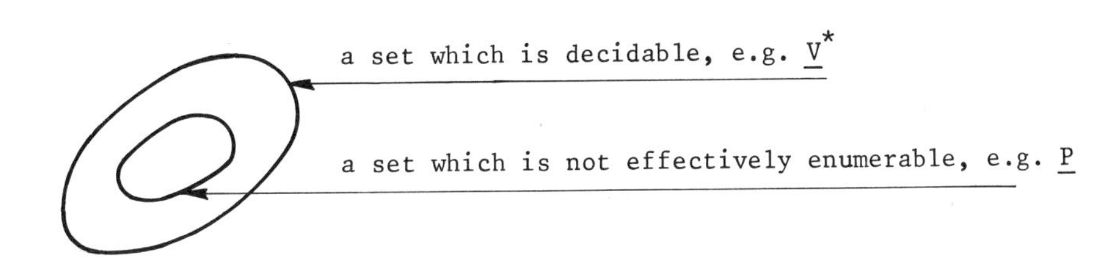

Note that this figure represents sets of strings whereas the figure of Section 6.8.2 represents sets of sets of strings.

Intuitively, this means that a decidable set differs from a set which is not decidable by the smaller "variety" of its strings rather than by the smaller "number" of its strings.

A similar remark holds for the effectively enumerable sets.

(*) The decision function of $\underline{V}^*$:

$$D : \underline{V}^* \to \{a, \varepsilon\} : D(x) = a$$

is clearly computable, hence $\underline{V}^*$ is decidable.

6.8.4. Axiomatic systems

An axiomatic system defines by construction an effectively enumerable set of strings. In mathematical logic, for instance, these strings are interpreted as theorems.

Gödel whose theorem was already mentioned in Section 3.1 has proved that the set of truths in mathematics is not effectively enumerable (while the set of truths in geometry is).

APPENDIX 1 : BIBLIOGRAPHICAL NOTES

As already mentioned in the Preface most treatises on computability are oriented towards the foundations of mathematics.

A classical and very clearly written book is [1]. Another recommendable volume is [2] but it is based on recursive functions rather than on Turing machines. For a good intuitive understanding [3] may be helpful. Note that there exists a great amount of other books devoted to the same subject.

A treatment of the subject somewhat more oriented towards computer science is to be found in some chapters of [4, 5, 6]: in [4] the subject is treated in a rigorous but very concise way; [5] insists on the intuitive understanding; [6] is in its goals related to the present notes but the study is conducted with less mathematical rigor.

[1] M. Davis, *Computability and Unsolvability*, McGraw-Hill, New York, 1958.

[2] Hartley Rogers, *Theory of Recursive Functions and Effective Computability*, McGraw-Hill, New York, 1967.

[3] M. Minsky, *Computation: Finite and Infinite Machines*, Prentice-Hall, Englewood Cliffs, 1967.

[4] M.A. Arbib, *Theories of Abstract Automata*, Prentice-Hall, Englewood Cliffs, 1969.

[5] M. Gross, A. Lentin, *Notions sur les grammaires formelles*, Gauthiers-Villars, Paris, 1967. (An English translation appeared in 1970).

[6] J.E. Hopcroft, J.D. Ullman, *Formal Languages and their Relation to Automata*, Addison-Wesley, Reading (Mass.), 1969.

APPENDIX 2 : LIST OF THE MOST IMPORTANT NOTATIONS

The Section Number indicates where the notation has been introduced.

(a, b, c)	1.2.2.1	$\overline{F}^{(i)}$	1.4.5.
{a, b, c}	1.2.2.2	ε	2.1
{ }	1.2.2.2	$\underline{V}$	2.1
$\{ x \mid P(x)\}$	1.2.2.2	$\underline{V}^*$	2.1
$a \in \underline{S}$	1.3.1	conc	2.2
$a \notin \underline{S}$	1.3.1	head	2.2
$\underline{S} \subseteq \underline{T}$	1.3.1	delete	2.2
$\underline{S} \subset \underline{T}$	1.3.1	reverse	2.2
card(n)	1.3.1	x^R	2.2
$\underline{S} \cup \underline{T}$	1.3.2	a^n	2.3
$\underline{S} \cap \underline{T}$	1.3.2	l(x)	2.3
$\underline{S} \setminus \underline{T}$	1.3.2	$\underline{S}.\underline{T}$	2.3
$\underline{S} - \underline{T}$	1.3.2	A	2.5
$\overline{\underline{S}}$	1.3.2	$(\underline{V}^{*n}, A)$-to-$\underline{N}$	2.7.1
$\underline{S} \times \underline{T}$	1.3.2	$\underline{V}^{*n}$-to-$\underline{N}$	2.7.1
$\underline{S}^n$	1.3.2	$\underline{V}^*$-to-$\underline{N}$	2.7.1
$\underline{D}_F$	1.4.2	$\underline{N}$-to-$(\underline{V}^*, A)$	2.7.1
$\underline{R}_F$	1.4.2	$\underline{N}$-to-$\underline{V}^*$	2.7.1
$\underline{N}$	1.4.2,1.5.2	$(\underline{V}^{*n}, A)$-to-$(\underline{W}^*, B)$	2.7.2
F	1.4.3	$\underline{V}^{*n}$-to-$\underline{W}^*$	2.7.2
$F(a_1, \ldots, a_n)$	1.4.3	$\underline{V}^*$-to-$\underline{W}^*$	2.7.2
$F : \underline{S} \rightarrow \underline{T}$	1.4.3	$\underline{Q}$	3.4.1
$F : \underline{S} \twoheadrightarrow \underline{T}$	1.4.3	B	3.4.1
$\overline{F}$	1.4.5	q_s	3.4.1

APPENDIX 3 : LIST OF THE MOST IMPORTANT CONCEPTS

The Section Number indicates where the concept has been introduced.

Lecture Notes in Economics and Mathematical Systems

(Vol. 1–15: Lecture Notes in Operations Research and Mathematical Economics, Vol. 16–59: Lecture Notes in Operations Research and Mathematical Systems)

Vol. 1: H. Bühlmann, H. Loeffel, E. Nievergelt, Einführung in die Theorie und Praxis der Entscheidung bei Unsicherheit. 2. Auflage, IV, 125 Seiten 4°. 1969. DM 16,–

Vol. 2: U. N. Bhat, A Study of the Queueing Systems M/G/1 and GI/M/1. VIII, 78 pages. 4°. 1968. DM 16,–

Vol. 3: A. Strauss, An Introduction to Optimal Control Theory. VI, 153 pages. 4°. 1968. DM 16,–

Vol. 4: Einführung in die Methode Branch and Bound. Herausgegeben von F. Weinberg. VIII, 159 Seiten. 4°. 1968. DM 16,–

Vol. 5: Hyvärinen, Information Theory for Systems Engineers. VIII, 205 pages. 4°. 1968. DM 16,–

Vol. 6: H. P. Künzi, O. Müller, E. Nievergelt, Einführungskursus in die dynamische Programmierung. IV, 103 Seiten. 4°. 1968. DM 16,–

Vol. 7: W. Popp, Einführung in die Theorie der Lagerhaltung. VI, 173 Seiten. 4°. 1968. DM 16,–

Vol. 8: J. Teghem, J. Loris-Teghem, J. P. Lambotte, Modèles d'Attente M/G/1 et GI/M/1 à Arrivées et Services en Groupes. IV, 53 pages. 4°. 1969. DM 16,–

Vol. 9: E. Schultze, Einführung in die mathematischen Grundlagen der Informationstheorie. VI, 116 Seiten. 4°. 1969. DM 16,–

Vol. 10: D. Hochstädter, Stochastische Lagerhaltungsmodelle. VI, 269 Seiten. 4°. 1969. DM 18,–

Vol. 11/12: Mathematical Systems Theory and Economics. Edited by H. W. Kuhn and G. P. Szegö. VIII, IV, 486 pages. 4°. 1969. DM 34,–

Vol. 13: Heuristische Planungsmethoden. Herausgegeben von F. Weinberg und C. A. Zehnder. II, 93 Seiten. 4°. 1969. DM 16,–

Vol. 14: Computing Methods in Optimization Problems. Edited by A. V. Balakrishnan. V, 191 pages. 4°. 1969. DM 16,–

Vol. 15: Economic Models, Estimation and Risk Programming: Essays in Honor of Gerhard Tintner. Edited by K. A. Fox, G. V. L. Narasimham and J. K. Sengupta. VIII, 461 pages. 4°. 1969. DM 24,–

Vol. 16: H. P. Künzi und W. Oettli, Nichtlineare Optimierung: Neuere Verfahren, Bibliographie. IV, 180 Seiten. 4°. 1969. DM 16,–

Vol. 17: H. Bauer und K. Neumann, Berechnung optimaler Steuerungen, Maximumprinzip und dynamische Optimierung. VIII, 188 Seiten. 4°. 1969. DM 16,–

Vol. 18: M. Wolff, Optimale Instandhaltungspolitiken in einfachen Systemen. V, 143 Seiten. 4°. 1970. DM 16,–

Vol. 19: L. Hyvärinen, Mathematical Modeling for Industrial Processes. VI, 122 pages. 4°. 1970. DM 16,–

Vol. 20: G. Uebe, Optimale Fahrpläne. IX, 161 Seiten. 4°. 1970. DM 16,–

Vol. 21: Th. Liebling, Graphentheorie in Planungs- und Tourenproblemen am Beispiel des städtischen Straßendienstes. IX, 118 Seiten. 4°. 1970. DM 16,–

Vol. 22: W. Eichhorn, Theorie der homogenen Produktionsfunktion. VIII, 119 Seiten. 4°. 1970. DM 16,–

Vol. 23: A. Ghosal, Some Aspects of Queueing and Storage Systems. IV, 93 pages. 4°. 1970. DM 16,–

Vol. 24: Feichtinger, Lernprozesse in stochastischen Automaten. V, 66 Seiten. 4°. 1970. DM 16,–

Vol. 25: R. Henn und O. Opitz, Konsum- und Produktionstheorie. I. II, 124 Seiten. 4°. 1970. DM 16,–

Vol. 26: D. Hochstädter und G. Uebe, Ökonometrische Methoden. XII, 250 Seiten. 4°. 1970. DM 18,–

Vol. 27: I. H. Mufti, Computational Methods in Optimal Control Problems. IV, 45 pages. 4°. 1970. DM 16,–

Vol. 28: Theoretical Approaches to Non-Numerical Problem Solving. Edited by R. B. Banerji and M. D. Mesarovic. VI, 466 pages. 4°. 1970. DM 24,–

Vol. 29: S. E. Elmaghraby, Some Network Models in Management Science. III, 177 pages. 4°. 1970. DM 16,–

Vol. 30: H. Noltemeier, Sensitivitätsanalyse bei diskreten linearen Optimierungsproblemen. VI, 102 Seiten. 4°. 1970. DM 16,–

Vol. 31: M. Kühlmeyer, Die nichtzentrale t-Verteilung. II, 106 Seiten. 4°. 1970. DM 16,–

Vol. 32: F. Bartholomes und G. Hotz, Homomorphismen und Reduktionen linearer Sprachen. XII, 143 Seiten. 4°. 1970. DM 16,–

Vol. 33: K. Hinderer, Foundations of Non-stationary Dynamic Programming with Discrete Time Parameter. VI, 160 pages. 4°. 1970. DM 16,–

Vol. 34: H. Störmer, Semi-Markoff-Prozesse mit endlich vielen Zuständen. Theorie und Anwendungen. VII, 128 Seiten. 4°. 1970. DM 16,–

Vol. 35: F. Ferschl, Markovketten. VI, 168 Seiten. 4°. 1970. DM 16,–

Vol. 36: M. P. J. Magill, On a General Economic Theory of Motion. VI, 95 pages. 4°. 1970. DM 16,–

Vol. 37: H. Müller-Merbach, On Round-Off Errors in Linear Programming. VI, 48 pages. 4°. 1970. DM 16,–

Vol. 38: Statistische Methoden I, herausgegeben von E. Walter. VIII, 338 Seiten. 4°. 1970. DM 22,–

Vol. 39: Statistische Methoden II, herausgegeben von E. Walter. IV, 155 Seiten. 4°. 1970. DM 16,–

Vol. 40: H. Drygas, The Coordinate-Free Approach to Gauss-Markov Estimation. VIII, 113 pages. 4°. 1970. DM 16,–

Vol. 41: U. Ueing, Zwei Lösungsmethoden für nichtkonvexe Programmierungsprobleme. VI, 92 Seiten. 4°. 1971. DM 16,–

Vol. 42: A. V. Balakrishnan, Introduction to Optimization Theory in a Hilbert Space. IV, 153 pages. 4°. 1971. DM 16,–

Vol. 43: J. A. Morales, Bayesian Full Information Structural Analysis. VI, 154 pages. 4°. 1971. DM 16,–

Vol. 44: G. Feichtinger, Stochastische Modelle demographischer Prozesse. XIII, 404 pages. 4°. 1971. DM 28,–

Vol. 45: K. Wendler, Hauptaustauschschritte (Principal Pivoting). II, 64 pages. 4°. 1971. DM 16,–

Vol. 46: C. Boucher, Leçons sur la théorie des automates mathématiques. VIII, 193 pages. 4°. 1971. DM 18,–

Vol. 47: H. A. Nour Eldin, Optimierung linearer Regelsysteme mit quadratischer Zielfunktion. VIII, 163 pages. 4°. 1971. DM 16,–

Vol. 48: M. Constam, Fortran für Anfänger. VI, 143 pages. 4°. 1971. DM 16,–

Vol. 49: Ch. Schneeweiß, Regelungstechnische stochastische Optimierungsverfahren. XI, 254 pages. 4°. 1971. DM 22,–

Vol. 50: Unternehmensforschung Heute – Übersichtsvorträge der Züricher Tagung von SVOR und DGU, September 1970. Herausgegeben von M. Beckmann. VI, 133 pages. 4°. 1971. DM 16,–

Vol. 51: Digitale Simulation. Herausgegeben von K. Bauknecht und W. Nef. IV, 207 pages. 4°. 1971. DM 18,–

Vol. 52: Invariant Imbedding. Proceedings of the Summer Workshop on Invariant Imbedding Held at the University of Southern California, June – August 1970. Edited by R. E. Bellman and E. D. Denman. IV, 148 pages. 4°. 1971. DM 16,–

Vol. 53: J. Rosenmüller, Kooperative Spiele und Märkte. IV, 152 pages. 4°. 1971. DM 16,–

Vol. 54: C. C. von Weizsäcker, Steady State Capital Theory. III, 102 pages. 4°. 1971. DM 16,–

Vol. 55: P. A. V. B. Swamy, Statistical Inference in Random Coefficient Regression Models. VIII, 209 pages. 4°. 1971. DM 20,–

Vol. 56: Mohamed A. El-Hodiri, Constrained Extrema. Introduction to the Differentiable Case with Economic Applications. III, 130 pages. 4°. 1971. DM 16,–

Vol. 57: E. Freund, Zeitvariable Mehrgrößensysteme. VII, 160 pages. 4°. 1971. DM 18,–

Vol. 58: P. B. Hagelschuer, Theorie der linearen Dekomposition. VII, 191 pages. 4°.1971. DM 18,–

Vol. 59: J. A. Hanson, Growth in Open Economics. IV, 127 pages. 4°. 1971. DM 16,–

Vol. 60: H. Hauptmann, Schätz- und Kontrolltheorie in stetigen dynamischen Wirtschaftsmodellen. V, 104 pages. 4°. 1971. DM 16,–

Vol. 61: K. H. F. Meyer, Wartesysteme mit variabler Bearbeitungsrate. VII, 314 pages. 4°. 1971. DM 24,–

Vol. 62: W. Krelle u. G. Gabisch unter Mitarbeit von J. Burgermeister, Wachstumstheorie. VII, 223 pages. 4°. 1972. DM 20,–

Vol. 63: J. Kohlas, Monte Carlo Simulation im Operations Research. VI, 162 pages. 4°. 1972. DM 16,–

Vol. 64: P. Gessner u. K. Spremann, Optimierung in Funktionenräumen. IV, 120 pages. 4°. 1972. DM 16,–.

Vol. 65: W. Everling, Exercises in Computer Systems Analysis. VIII, 184 pages. 4°. 1972. DM 18,–

Vol. 66: F. Bauer, P. Garabedian und D. Korn, Supercritical Wing Sections. V, 211 pages. 4°. 1972. DM 20,–.

Vol. 67: I. V. Girsanov, Lectures on Mathematical Theory of Extremum Problems. V, 136 pages. 4°. 1972. DM 16,–.

Vol. 68: J. Loeckx, Computability and Decidability. An Introduction for Students of Computer Science. VI, 76 pages. 4°. 1972. DM 16,–